A Continent Transformed

Other Meridian titles:

Robert H Fagan and Michael Webber, GLOBAL RESTRUCTURING

A CONTINENT TRANSFORMED

Human Impact on the Natural Vegetation of Australia

JAMIE KIRKPATRICK

AUSTRALIAN
GEOGRAPHICAL
PERSPECTIVES

Series editors
Deirdre Dragovich
Alaric Maude

Melbourne
OXFORD UNIVERSITY PRESS
Oxford Auckland New York

OXFORD UNIVERSITY PRESS AUSTRALIA
Oxford New York Toronto
Delhi Bombay Calcutta Madras Karachi
Kuala Lumpur Singapore Hong Kong Tokyo
Nairobi Dar es Salaam Cape Town
Melbourne Auckland Madrid
and associated companies in
Berlin Ibadan

OXFORD is a trade mark of Oxford University Press

© Jamie Kirkpatrick 1994
First published 1994

This book is copyright. Apart from any fair
dealing for the purposes of private study,
research, criticism or review as permitted under
the Copyright Act, no part may be reproduced,
stored in a retrieval system, or transmitted, in
any form or by any means, electronic, mechanical,
photocopying, recording or otherwise without
prior written permission. Enquiries to be made to
Oxford University Press.

Copying for educational purposes
Where copies of part or the whole of the book are
made under Section 53B or Section 53D of the Act,
the law requires that records of such copying be
kept. In such cases the copyright owner is
entitled to claim payment.

National Library of Australia
Cataloguing-in-Publication data:

Kirkpatrick, J. B. (James Barrie).
A continent transformed.

Bibliography.
Includes index.
ISBN 0 19 553473 5.

1. Vegetation dynamics — Australia. 2. Natural resources —
Australia. 3. Human ecology — Australia. 4. Man — Influence
on nature — Australia. I. Institute of Australian Geographers.
II. Title. (Series : Meridian, Australian geographical
perspectives).

581.52640994

Edited by Jenny Missen
Cover photograph from Animal and
Plant Control Commission, South Australia
Typeset by SRM, Malaysia
Printed by SRM, Malaysia
Published by Oxford University Press,
253 Normanby Road, South Melbourne, Australia

Foreword

Australian geographers have produced some excellent books in recent years, several of them in association with the 1988 bicentennial of European settlement in the continent, and all of them building on the maturing of geographical research in this country. However, there is a continuing need for relatively short, low cost books written for university students to fill the gap between chapter-length surveys and full-length books, to explore the geographical issues and problems of Australia and its region, or to present an Australian perspective on global geographical processes.

Meridian: Australian Geographical Perspectives is a series initiated by the Institute of Australian Geographers to fill this need. The term meridian refers to a line of longitude linking points in a half-circle between the poles. In this series it symbolises the interconnections between places in the global environment and global economy, which is one of the Key themes of contemporary geography. The books in the series cover a variety of physical, environmental, economic and social geography topics, and are written for use in first and second year courses where the existing texts and reference books lack a significant Australian perspective. To cope with the very varied content of geography courses taught in Australian universities the books are designed not as comprehensive texts, but as modules on specific themes which can be used in a variety of courses. They are intended to be used either in a one-semester course, or in a one-semester component of a full year course.

Titles in the series will cover a range of topics representing contemporary Australian geographical teaching and research, such as economic restructuring, vegetation change, natural hazards, the changing nature of cities, land degradation, gender and geography, and urban environmental problems. Although the emphasis in the series is on Australia, we also intend to produce some titles on Southeast Asia, using the considerable expertise that some Australian geographers have developed on this region.

We hope geography students will find the series informative, lively and relevant to their interests. Individual titles will also be of interest to students in related disciplines, such as environmental sciences, planning, economics, women's studies and Asian Studies.

While the primary aim of the series is to produce books for students, the topics selected deal with issues of relevance to all Australians. We therefore hope that the general reader will find some of the titles of interest, and discover that geographers have something distinctive to say on contemporary environmental, economic and social issues. As the books assume little or no previous training in geography, and attempt to avoid a textbook style, they should be readily understood by the general reader.

Jamie Kirkpatrick's book, *A Continent Transformed — Human Impact on the Natural Vegetation of Australia*, is the first in the series. Professor Kirkpatrick is one of Australia's most active and experienced plant geographers and conservation ecologists, and his book examines the changes in the vegetation cover of Australia produced by both the Aboriginal and European inhabitants of the continent. The loss or transformation of this vegetation cover, and of the plant communities represented in it, has created some major environmental problems for Australians. Jamie Kirkpatrick explores some of the ecological problems to be solved in achieving conservation of Australia's native vegetation, and suggests some solutions. His book presents contemporary scientific understanding of these issues, based on the disciplines of biogeography and ecology, in a very readable and personal account of an important topic.

Deirdre Dragovich
University of Sydney

Alaric Maude
Flinders University

Contents

1 Introduction — 1
The importance of biological diversity — 1
   The rights of the rest of the living world — 1
   Spiritual values — 2
   Ecological services — 2
   Economic worth — 3
The importance of Australia's biodiversity — 3
Ecologically sustainable development — 3
Some comments on science and plant geography/ecology — 4
About this book — 5

2 Understanding the distribution of species and ecosystems — 7
Species and genotypes — 7
Evolution and extinction — 8
Describing the distribution of species — 11
Explaining the distribution of species — 11
   Limits of tolerance — 11
   The regeneration niche and vegetation dynamics — 17

Dispersal and migration	19
Ecosystems and communities	20
Richness and diversity	22
An integrated explanation of species distribution patterns	23
A concluding preamble	24

3 Glaciers and Aborigines 25

Prehuman environmental history	25
Sources of evidence on the prehistoric human impact	26
The arrival of *Homo sapiens*	28
The nature of the prehistoric human impact	30
People in harmony with nature?	34
Current Aboriginal interaction with the environment	35
Another concluding preamble	36

4 Bush destruction and the creation of cultural vegetation 37

The incidence and magnitude of European bush destruction	37
Bush resistance to clearance	40
The transformation of the plains	41
Technology and bush destruction	43
Tree farming	45
Urbanisation	47
The characteristics and ecology of Australian cultural vegetation	48
The convergence of clearance and reservation	50

5 The impact of forest use 53

A very brief history of wood extraction	54
The forestry paradigm	55
The impact of silvicultural systems on eucalypt forest	55
The impact of rainforest logging	62
Impact on forests of the wet/dry tropics	65
Some ecological arguments related to forest logging	66
Conclusion	67

6 Fire 68

A brief history of post-invasion fire regimes 68
Fire-susceptible vegetation 69
Fire management in rainforest and alpine vegetation 71
Fuel reduction burning 72
Seasonality of burns 74
The interaction between fire and grazing 75
Interaction of fire with exotic plants 79
Fire management in protected areas 79

7 The invaders 82

A brief history of introductions 83
Is disturbance necessary for exotic invasion? 84
Characteristics and distribution 84
Eutrophication and invasion 85
Time bombs 87
Suburban and farm escapees 88
Control of environmental weeds 89
Our most threatening plant invader—the Cinnamon Fungus 92
The impact of invading animals 95
Our most threatening animal invader—the rabbit 97
Conclusion 98

8 Conserving the bush 99

Criteria for protected area selection 100
Size of reserves 103
   Minimum population size 103
   The theory of island biogeography 105
   Evolutionary continuity 106
   The problem of edge effects 108
Location of reserves 109
People management problems 110
A synopsis of the state of various vegetation types 116
   Rainforest 116
   Alpine vegetation 116
   Wet eucalypt forest 116
   Dry eucalypt forest 117

    Tropical eucalypt forest — 117
    Temperate grassy woodland and grassland — 118
    Mallee — 118
    Mitchell (Astrebla) grasslands — 118
    Heath — 119
    Desert vegetation — 119
    Wetlands — 119
    Coastal vegetation — 119
  Integrated land management — 120

References and further reading — 121

Index — 126

1

Introduction

THE IMPORTANCE OF BIOLOGICAL DIVERSITY

The variety of life gives sustenance, shelter and a variety of experiences to all our lives. Biological diversity, sometimes called biodiversity, consists of species and their genetic variants, and of the assemblages of species we call communities. There are many good reasons for maintaining the biodiversity of our planet, including that part of it that is the subject of this book, the terrestrial vegetation of Australia:

The rights of the rest of the living world

Many philosophers believe that the living world has intrinsic value, independent of the existence and views of human beings. In a strict sense, rights can only result from human ethical judgements expressed through law and custom. However, we are accustomed to groups of people labelling a particular moral outcome that they favour as a 'right'. The argument for a moral right to continued existence for all genotypes, species and communities often rests on a perception of the communality of life. The extension of our perception of self to envelop the biosphere has been argued to be a prerequisite for planetary health.

Adopting the viewpoint that all species have a right to exist does not necessarily imply that all individuals of all species have a right to exist. In fact, there is often conflict between biodiversity conservation and animal rights, as in the well-publicised case of the shooting of kangaroos in Hattah-Kulkyne National Park. This shooting of individuals of a common animal species was undertaken to restore the pre-grazing diversity of habitat.

Conflict is more intense with human rights. Many opponents of the intrinsic value of all of the variety of life see the survival of human individuals, or even an increase in their net material welfare, to be of greater moral importance than the maintenance of elements of biodiversity that they consider to be redundant, in that they do not contribute to the direct satisfaction of human needs. Most religions have adopted this philosophical viewpoint.

Spiritual values

Biodiversity provides many benefits to people that are not amenable to economic evaluation. People may value the richness of life for the continuity it provides between the past and the future, for its beauty, for its curiosity value, for its scientific interest or for its integral role in their culture. In other words, people may wish to maintain biodiversity because they feel happier that way. Of course, many others are barely aware of the existence of other species, and many people who are aware are indifferent to the fate of our non-human relatives. Some people see most of nature as an enemy awaiting eradication.

Ecological services

Many of the best things in life are only free courtesy of the service functions of biodiversity. The air is only rich in oxygen because of a long history of respiration by plants. Streams only have a constant flow because vegetation protects and helps develop soils that act as a sponge, providing a reservoir. Water flows fresh only because catchment forests soak up the moisture that would otherwise flush out the salt. Paddocks are not buried knee deep in cow dung because animals and plants make a living from decomposition. Of course, one species of dung beetle may do the job, and plantations may prevent salinisation as effectively as native forests. The instrumental approach to nature recognises redundancy. The problem is knowing exactly what is redundant as a tool or resource. Should species be allowed to become extinct because we have no evidence that they are necessary for maintaining the ecosystems that support humanity, or, should biodiversity be protected because we do not know whether it is useful for people or not? The former approach minimises the inconvenience of biodiversity maintenance. The latter approach minimises the risks to the service systems of the planet.

Economic worth

Humanity uses a remarkably small proportion of global biodiversity for direct economic gain. However, we would all be dead without it, and the potential for economic gain from other organisms is considerable. For example, some rare Australian native species in the genus *Glycine* are perennials. The genes for their longevity may be valuable in improving another *Glycine*, the soy-bean. The chemicals found in plants are used widely in pharmacology. Most Australian plants have yet to be investigated to this end.

Tourism is now one of Australia's major industries. A large proportion of the $20–25 billion a year contributed by this industry rests on the survival of Australia's biota in the wild. Natural pastures form the basis for much of Australia's primary production. Natural forests are used for wood production, honey production and for water catchments. Australia's wildflowers are used widely in the horticultural industry.

Thus our lives, our environment, our recreation and our spiritual harmony all ultimately rest on the foundation of biological diversity. Its maintenance therefore becomes one of the great human problems, in more ways than one. How then does Australia perform as a container and maintainer of biodiversity?

THE IMPORTANCE OF AUSTRALIA'S BIODIVERSITY

Australia is a megabiodiversity country. Our 25,000 higher plant species constitute approximately 10% of the global total, and most of these species are found only in Australia. We are the only large, rich country with a substantial area of land in the tropics. The brief period of occupance by an agricultural and industrial society has yet to completely eliminate the landscapes that arose from the interaction of gatherers and hunters with Australian vegetation. However, the consequences of the European invasion on biodiversity have been substantial, including the almost total destruction of natural (pre-invasion) ecosystems, such as the temperate grasslands, and the extinction of a large number of animals and plants.

ECOLOGICALLY SUSTAINABLE DEVELOPMENT

While the processes of natural vegetation loss and species extinction continue to rework the face of our continent, there has been a shift in the political perception of the importance of biodiversity. This has been expressed most cogently in the wide consensus on the desirability of ecologically sustainable development. The World Commission on Environment and Development produced a strategy, known as the Brundtland report, that has gained wide international acceptance for a concept of sustainable development that married ongoing global economic growth with environmental protection. This positive response was influenced by the ambiguous nature of the concept, and its acceptance of ongoing economic growth. Sustainable development is a markedly instrumental approach to conservation, with a firm anthropocentric focus. Thus, it accepts biodiversity loss in the cause of human welfare.

The conservation groups in Australia convinced the Commonwealth Government that its exploration of the implications of sustainable development should employ the qualifier 'ecologically'. Thus, the 1990–91 discussions of the Australian Ecologically Sustainable Development Working Groups developed recommendations for the future under the trinity of economic development, maintenance of landscape productivity and maintenance of biological diversity. In this context, the meaning of biodiversity maintenance was the survival of genotypes, species and communities at a world level. The Australian Conservation Foundation has argued that a local level of biodiversity maintenance is more appropriate. Otherwise, a species could be destroyed over most of its range without necessarily violating the national goal.

Some comments on science and plant geography/ecology

Science is the subdiscipline of philosophy concerned with the refutation of testable hypotheses. Hypotheses are formulated within the context of over-riding theories which set the guidelines for inquiry. Theories and hypotheses are never proven, they are simply not refuted. Several different theoretical constructs may satisfy the data available. Therefore, there is a subjective element in the choice of theory. There is also a subjective element in the choice of an hypothesis for testing. There are a limited number of scientists and a limitless number of hypotheses. The values of society and, sometimes, the investigator determine which hypotheses will be tested. In many cases, especially in plant geography and ecology, research is a response to a vague curiosity about a natural phenomenon, an area of vegetation, or a relationship between plants and environment, without any specific articulation of hypotheses or direct recourse to theory.

Science is largely a reductionist procedure, that is, the nature of a whole is approached by the study of its component parts. This has, somewhat unjustly, been compared to trying to understand a symphony by studying the individual notes in the score. Ecology, in common understanding, is holistic, but its scientific reality is largely reductionist (see, for example, papers in the *Australian Journal of Ecology*). This emphasis on researching the parts rather than the whole means that the conclusions from studies cannot necessarily be used in the real world. For example, one study might show that a particular plant species can withstand fire. Another study might show that the same plant can survive heavy grazing. Managers might therefore think that they could graze and burn the plant without any fear of its loss — yet the combination of burning followed by grazing might lead to its extinction.

After several decades of intensive scientific investigation of the ecology of Australia we know sufficient about many ecosystems to be able to predict some of their short and medium-term responses to some forms of disturbance regimes, but we cannot be confident of outcomes in the long term, and seem incapable of predicting the initiation of major ecological perturbations, such as the dieback of native species from root rot fungi. Our lack of predictive ability relates partly to a lack of scientific investigation in the reductionist mode, partly to the chaotic (inherently unpredictable) behaviour of ecological systems at particular time and space scales, and partly to the existence of ecological processes that are synergistic in their nature, that is, the sum of individual effects is different from the sum of interacting effects.

The absence of fully certified scientific truth and the essentially sociological basis of science may induce insecurity in the casual observer, but the accumulation of observations, experimental results and long-term monitoring data that is taking place in Australia today means that we are more and more likely to be able to make non-disastrous vegetation management decisions, irrespective of the paradigm (view of the world) in which these data are placed, or the sociological context in which they were derived.

ABOUT THIS BOOK

This book documents and dissects the rapid changes that have taken place, are taking place and probably will take place in Australian terrestrial vegetation. Chapter 2 provides the ecological vocabulary and conceptual tools that form the basis of a scientific understanding of pattern and process in the Australian native biota. The role of hunters and gatherers and their descendants in modifying Australian ecosystems forms the focus of the third chapter. The nature and consequences of the activities of Europeans in Australia over the last 200 years are explored in the subsequent chapters, which cover land clearance and the creation of cultural vegetation, forest modification, the role of fire and the impact of introduced animals and plants. The final chapter covers the science of native plant biodiversity maintenance in Australia. In all these areas there are considerable divergences of opinion. In some cases these relate to the scientific interpretation of patchy and incomplete information; in others they relate to the value frameworks of the protagonists. I have attempted to clarify the nature of these differences, while avoiding any pretence of total objectivity (see above).

2

Understanding the distribution of species and ecosystems

SPECIES AND GENOTYPES

General classifications aim to erect classes which enable the maximum number of significant generalisations to be made about any individual in that class. The basic general classificatory unit for living things is the species. Each species is described by a scientific binomial as well as by a common name, for example, *Eucalyptus globulus* (Tasmanian Blue Gum). Species are aggregated into genera (e.g. *Eucalyptus*), which in turn are aggregated into families (e.g. the genus *Eucalyptus* belongs to the family Myrtaceae). Scientific binomials should be distinguished by underlining or italicising. The generic name always begins with a capital letter and the species name always begins with a lower case letter, even if it is someone's name.

Taxonomists describe plant species after examining specimens that have been collected and lodged in herbaria (sing. herbarium). They group specimens according to their similarity, placing most emphasis on those characteristics, such as those of the reproductive parts, that are least influenced by environmental conditions. They then search for consistent differences between these groups. For example, there might be a morphological discontinuity in fruit size, such that one species always has fruits larger than 20 mm in diameter, while the other always has fruits less than 15 mm in diameter. In animal taxonomy an attempt is made to ensure that different species are incapable of crossing and producing fertile offspring. In plant taxonomy such a rule would produce an unworkable result. For example, the tallest flowering plant in the world, *Eucalyptus regnans*, and the morphologically and ecologically distinct small tree, *Eucalyptus tenuiramis*, are linked through a chain of hybridising species. Morphological discontinuities, or even partial discontinuities, are sufficient for plant species recognition as long as there are reasonable grounds for believing that such discontinuities are genetically based.

To gain scientific recognition for a previously undescribed species, a taxonomist is required to write a description in Latin, have that description published and to designate, and lodge in an appropriate institution, a type specimen, which acts as a reference point. The scientific names of species are subject to change, but at any one time they are usually accepted by all users. This lack of stability is caused by:

- the strong element of judgement required in determining taxonomic boundaries — taxonomists can be either lumpers (grouping species) or splitters (dividing species) at all levels of the hierarchy, and lumpers often alternate with splitters;
- increased knowledge of variation as a result of further collecting, or improved measurement of characteristics;
- the rectification of grammatical mistakes in the original binomial (e.g. *Eucalyptus sieberiana* was grammatically incorrect and was changed to *Eucalyptus sieberi*);
- the discovery of earlier legal descriptions of species published during the species description races of the eighteenth and nineteenth centuries — e.g. the name *Eucalyptus gigantea* was published later than the name *Eucalyptus delegatensis* for the same taxon (taxonomic unit, pl. taxa).

While scientific names are temporally unstable, it is possible to follow a particular source for the nomenclature of many groups of organisms. For example, there is a recently published census of Australian plants which cross-references old names and new. It is certainly better to use scientific names than common names, which change from place to place as well as over time, and cover far fewer species.

The above discussion might give the impression that species are fairly arbitrarily defined. In fact, most species have remained stable in concept, if not name, since their first recognition, due to their readily perceptible distinctiveness and uniformity.

EVOLUTION AND EXTINCTION

Most scientists who are not fundamentalist Christians believe species to be the product of biological evolution. Mutation and sexual recombination provide the raw material of variation that is worked upon by selective processes. Mutation occurs when an external force, such as radiation, changes the nature of heritable characteristics. Most mutations are negative in their impact, making the organism that results less fit to survive and reproduce. A few might confer a reproductive advantage in particular environmental situations. If individuals with mutated genotypes produce more reproducing progeny than those with the old pattern, they will eventually displace the old pattern. Sexual recombination gives an enormous number of options for selection in relation to environmental cues. Recently, some scientists have suggested that some organisms may have derived some characteristics as a result of viral modifications of their genes. This is akin to the results of genetic engineering, and could be an explanation of some shared characteristics between otherwise taxonomically disparate species.

There are two main explanations of the process of speciation, which is the development of morphological discontinuities and breeding barriers that ultimately result in the recognition of a new species. ALLOPATRIC SPECIATION is thought to occur where sets of populations (a population is a group of potentially interbreeding individuals) become isolated by environmental change. As all environments are different in some way, divergence occurs in response to selective pressures. Divergence may also occur as a result of chance events. PARAPATRIC SPECIATION takes place within continuous populations where an environmental break, such as a major soil boundary, occurs. The progeny of individuals that cross with individuals on the other side of the boundary are likely to be less successful than those of individuals that cannot cross in this way. Thus, selection will result in the formation of breeding barriers, unless the occasional selective advantages of such crossing outweigh the selective disadvantages. This latter situation seems to occur in some plants that produce enormous numbers of progeny that are doomed to density-dependent mortality before reproduction takes place, simply because there is no space for all the individuals to grow to maturity. The many unfit individuals in a progeny will die anyway. The few that might have gained benefit from the over-boundary outcrossing will survive. It is theoretically also possible that speciation could occur sympatrically, that is within a population on the one site. This would require divergent selective pressures within that site.

As already indicated, selective pressures do not necessarily result in speciation. Most species that have been studied have been shown to exhibit geographic variation in at least some of their genetically-determined characteristics, and this geographic variation can usually be related to variation in environment. This variation is clinal, or more or less continuous, where the critical environmental gradient is continuous. It can be ecotypic, or patchy, where the critical environmental factor is discontinuous in its distribution. Thus, clinal variation tends to occur in response to environmental factors such as moisture availability and summer warmth, while ecotypic variation occurs in response to environmental factors such as soil type. Any species may exhibit clinal variation in some characters and ecotypic variation in others.

Hybrids are the product of crossing between recognised species. HYBRIDISATION between plant species is a common phenomenon easily confused with geographic variation. However, populations of hybrids tend to be much more variable than clinal populations. The model for hybridisation is isolation, partial divergence and recontact. However, the symptoms of hybridisation would also be apparent in the last stages of parapatric speciation. Hybrids usually only survive to reproduce in the narrow intermediate environment of the contiguous margins of the parent species' distributions. If disturbance in this area creates environmental heterogeneity, hybrid swarms may result. In changing environmental conditions a hybrid genotype may be fitter than either parent, and stabilising selection then may result in a new species. Populations on the margins of the distribution of a species are often subject to the greatest selective pressures, and are most likely to be the sources of new species through hybridisation and stabilising selection.

While new species evolve others become extinct. Although extinctions tend to occur in pulses associated with dramatic environmental changes, such as the impact of the meteor which some scientists think was responsible for the loss of the dinosaurs, there is a continual loss of species related to less spectacular causes. Most species are relatively uncommon and highly specialised. The chance fluctuations of environmental conditions, or the evolution of a new pathogen (disease organism) or predator, may cause extinction without any major or unusual environmental perturbation. The natural long-term extinction rate has been calculated to be in the order of 0.02–2 species per annum. The global extinction rates have been well in excess of this figure over the last few hundred years, and Australia's record has been particularly bad. However, some of the figures given for actual and prospective species loss in the latter half of the twentieth century seem excessive given that Australia, with the worst record for plant extinctions over the last 200 years of any country in the world, has lost only 90–100 higher plant species out of an approximate total of 25,000. Most of the high range of figures are based on the unproven and unlikely assumption that the enormous number of invertebrate species associated with rainforest canopies have highly limited distributions. There is no doubt that the total elimination of any ecosystem will result in massive extinction. However, the effect of a reduction in area of an ecosystem is more problematical. For example, we know that rainforest in Australia is much more extensive today, even after massive clearing, than during the colder periods of the last 1.8 million years (the Quaternary). Thus, most rainforest species are reasonably secure despite massive reductions in their ranges. However, rainforest extinctions and endangerment have resulted from almost all of the range of particular rainforest types being cleared.

Describing the distribution of species

A fundamental characteristic of any species is that it has a distribution. With a few species this distribution may be continuous, that is, there is no gap within the distribution that is wider than the normal ability of individuals to transport themselves or have their propagules (e.g. seed, fruit) or themselves transported. Where such gaps exist, a species is said to have a disjunct distribution (the gaps are called disjunctions). In Australia most species have disjunctions within their range.

Species with a range that is severely contracted from a past distribution are known as RELICTS or relic species. Most Australian rainforest species would fall within this class, as the range of their habitat has dramatically contracted over the last few million years. However, we lack hard fossil data for the past distributions of most species, so the label relict is often conjectural.

Species that are confined to a particular area are said to be ENDEMIC to that area. Thus, a Kangaroo Island endemic species would only be known to occur naturally on Kangaroo Island. For a species to be considered an Australian native, the balance of evidence must suggest that it was present in Australia before the European invasion. The adjectives 'introduced', 'alien' and 'exotic' pertain to those species that have naturalised, that is, established reproducing populations in Australia, since that invasion. Exotics are usually species introduced from outside Australia, but can also be native species that have established reproducing populations outside their natural range as a result of recent human activity.

Explaining the distribution of species

Limits of tolerance

Species need certain substances in certain amounts over certain time periods to be able to survive and reproduce. Thus, most plants require solar radiation in the photosynthetic wavelength range, heat, water and a wide range of mineral nutrients, the most important of which are nitrogen and phosphorus. Each species has both a lower and upper limit in relation to the building blocks of life. For example, *Acacia suaveolens* requires some phosphorus to survive, but will die if phosphorus attains the levels usually found in domestic gardens. The potential range of a species is controlled by those factors that are closest to limiting by their deficiency or excess. The critical factors usually vary in space. For example, the upper altitudinal limit of *Eucalyptus delegatensis* is likely to relate to insufficient summer heat, while the lower altitudinal limit may be caused by insufficiencies of moisture. In relation to any one factor, species tend to perform best near the middle of their range, with their performance attenuating at the extremes. These ecological response curves are often bell-shaped, although they are not necessarily perfectly symmetrical. Different species tend to have different response curves for at least some of the various necessities of life.

The water factor

Moisture availability is one of the major controls on the distributions of terrestrial organisms in Australia. Terrestrial organisms are largely composed of water. They have the problem of ensuring that the water they lose through their cuticles, respiration and the operation of their cooling systems does not exceed the amount they take in through their root systems. Australian plants have developed an awe-inspiring range of adaptations that mitigate the water balance problem. They have developed:

- the ability to chemically fix carbon dioxide at night while keeping stomata closed during the day (crassulacean or C4 metabolism, e.g. *Crassula*);
- modifications to internal structure that allow rehydration after total desiccation (resurrection plants, e.g. *Borya*);
- the ability to store water in leaves, stems or roots (e.g. mallee eucalypt roots);
- photosynthesising stems (cladodes, e.g. *Allocasuarina*) or leaf stalks (phyllodes; e.g. *Acacia*);
- the ability to close stomata under water stress (e.g. dry country eucalypts);
- massive, deep root systems (e.g. many desert shrubs);
- hairs, wax and grooves that minimise the removal of moist air around stomata; and
- the ability to complete their life cycle while moisture is in plentiful supply (annuals).

Adaptations to avoid water stress usually impose a substantial growth cost. Thus, species that occur in well-watered areas are much more susceptible to drought than those in dry areas. For example, wet country eucalypts are not adept at closing their stomata in response to soil moisture deficits, so will die in the same conditions that allow the survival of dry country eucalypts. However, the superior potential growth rates of the wet country eucalypts will cause them to out-compete the dry country eucalypts in environments where water is not limiting.

Drought is a relative term. A drought for a rainforest tree might be two weeks without rain whereas a drought for a desert shrub might be two years. Drought will lead to major mortality of individuals of many species in the driest parts of their ranges. However, it rarely seems to directly cause major changes in the distributions of species, largely because any such adjustments would have taken place in the past. Also, the differences in soil moisture availability between north and south-facing slopes can be equivalent to many tens of millimetres of rainfall per year, because of the effects of slope and aspect on insolation and, therefore, evapotranspiration. While such differences are at their maximum in the temperate zone they are also strong in the wet/dry tropics where the sun is at its lowest during the dry season. Thus, species that might suffer severe mortality on north-facing slopes are likely to survive on nearby south-facing slopes.

The temperature factor

Species differ markedly in their responses to air and soil temperatures, these differences being the major reason for the marked latitudinal and altitudinal shifts in species composition that are characteristic of Australian ecosystems. Particular temperature ranges and durations are necessary signals for plant germination, plant flowering and plant growth. Extreme low temperatures can be a direct cause of widespread mortality. The most spectacular examples of this phenomenon have occurred on clear nights when extremely cold air flows to fill valleys. The line of the inversion layer can be clearly seen by the contrast of live and dead foliage. However, such frost-kill of foliage appears to have little effect on the long-term macrodistribution of species, although it may be part of the explanation for the treelessness of some subalpine and montane valleys.

The amount of heat seems more important in explaining species distributions than the degree of winter cold. The climatic treeline is the classical example of this phenomenon, being located where the mean temperature of the warmest month approximates 10°C. Air and soil temperatures rarely reach such heights that they are directly lethal to plant species. However, the desiccating effect of heat can be fatal, and high temperatures may encourage predators, parasites and competitors. Constantly high temperatures, such as those experienced in the tropics, prevent the growth and reproduction of species that require a cold period to break dormancy or initiate germination or reproduction.

The light factor

The vast majority of plants depend on solar radiation for their existence. Many plant species use variations in day length as signals to initiate or cease life processes. The distributions of plants may be limited through lack of light, usually as a result of shading by other plants. However, it is very easy to confuse the effects of lack of light with the other effects of a canopy, such as root competition for moisture and nutrients, and high humidity. Each individual photosynthesising plant has a light compensation curve, and different genotypes and species have different families of curves. At low and high light levels carbon fixation through photosynthesis is exceeded by respiration losses, forming the extremes of an usually positively skewed bell-shaped curve. The plants that perform best in shady conditions, especially if they are warm, have large leaves that optimise the capture of flecks and diffuse fields of solar radiation, and are slow-growing. The dominant genus in most of Australia's forests, *Eucalyptus*, is unusual in that the light that penetrates its vertical-leaved foliage is sufficient for successful growth of almost all other plants. Thus, eucalypt forests usually have dense understories, in contrast to the sparseness of the rainforest floor.

Solar radiation may become damaging to both plants and animals if too much is received in the shortest wavelengths. Areas at high altitude and low latitude receive the most ultraviolet solar radiation. The depletion of the ozone layer is allowing greater intake of short-wave radiation, especially at high latitudes. This may ultimately lead to changes in species' ranges.

The atmospheric factor

Most organisms have more than sufficient access to carbon dioxide and oxygen. These substances are seldom limiting. However, deoxygenated soils require special adaptations in plants: aerial roots, as in some mangrove species; breathing bark, as in some *Melaleuca* species; and the ability to reaerate soil, as in many wet heath species. Increased carbon dioxide content in the atmosphere seems likely to increase the growth rates of many plants, independent of any effect on temperatures.

The fluid nature of the atmosphere is important in influencing the distributions of species, particularly near the coast, where salt is carried, in the arid and semiarid zones where sand is carried, and in the high mountains where ice is carried. Species vary in their ability to withstand salt desiccation of their leaves, the abrasive affects of sand and ice and the drying effects of strong winds. Thus vegetation variation is often orthogonal (at right angles) to the prevailing wind direction.

The edaphic (soil) factor

The soil provides anchorage, water and nutrients to most plants. The ranges of many plant species are limited by the availability or excess of macro or micronutrients. Some of the major vegetation gradients in Australia are related closely to soil nutrient status. For example, the gradient from grassy forest to forests with understories dominated by small and hard-leaved (sclceromorphic) shrubs and sedges corresponds largely to the gradient from soils of high to low phosphorus content. Soils rich in calcium carbonate (lime) have dramatically different vegetation from those lacking this substance. Plants unadapted to lime suffer chlorosis (yellowing) in its presence, while those unadapted to lack of lime suffer dwarfing in its absence.

Soil type is critical in determining the amount and duration of moisture available to plants. Although clay soils potentially hold more water per unit volume than sandy soils they usually provide less water for plants. Low infiltration rates into clay soils, a higher proportion of water in the capillary fringe and limitations on root penetration related to poor aeration, all make clay soils less hospitable to continuously photosynthesising perennials than sandy soils, anywhere that potential evaporation exceeds precipitation. Thus, while in high rainfall areas the vegetation with the highest biomass is likely to be on clay soils, the reverse pertains to low rainfall areas.

The fire factor

The high temperatures associated with fire lead to the hyperthermal death of tissues and often result in the death of whole plants. Plant species differ dramatically in their response to single fires of varying intensity and to fire regimes. Many plants, such as orchids, lilies and bracken, survive all fires because they have underground organs that can reestablish the whole plant. Other plants, such as some wattles and native peas, deposit huge amounts of seed in the soil. This seed requires heat or mechanical disturbance for germination. Such heat is provided by the fire that kills the parent plant. Some species, such as Pigface and some rainforest trees, are killed by fire but resist its incidence through high foliage water or ash content. Species such as the cypress pines hold vast quantities of seed in cones that protect them against the heat of fire, although the parent plant is killed. Such species are susceptible to local extinction if any of the fire-free intervals in the fire regime are less than the period between germination and ripe cone production. Some species, such as most eucalypts, have so many mechanisms for recovery from fire as an individual or species that they are impervious to its impact. Others, such as the Deciduous Beech (*Nothofagus gunnii*) have no mechanisms at all. Species with all the fire-resistance syndromes listed above can be found growing close together in many areas of Australia, providing the challenge to the manager discussed in chapter 6.

The crux of the impact of fire on the distributions of plants in Australia is that fire-susceptible species, like Deciduous Beech, have ranges well within their climatic-edaphic potential, whereas fire-resistant species, like spinifex (*Triodia* and *Plectrachne*) have a range extended by fire to places in which they would otherwise be absent because of the effects of competition from other plants.

The biotic factor

Few plant species occupy the whole of their potential range. Competition from other plant species is the main reason for this restriction. Plants compete for light, moisture, nutrients and space. Where none of these factors is limiting there is usually a competition in which the most rapid growers gain the prize. These are known as RUDERALS. Where the physical factors are in abundant supply, but space is lacking, the glittering prize of survival goes to the species with a competitive strategy. These species are long-lived, space-possessing, relatively slow-growing and are capable of regeneration without the massive disturbance that favours the ruderals. Where light, moisture or nutrients are in short supply, the species that best tolerates their particular deficiency is likely to prevail. Unsurprisingly, they have been termed TOLERATORS. However, even ruderals and tolerators compete. For example, cliff species have to be tolerant of limited moisture and nutrients, but compete for the few cracks and ledges on which soil can accumulate.

Predation is another major influence on the distribution of plant species. This activity involves the physical removal of a part of an individual of one species to the benefit of another species. For example, Koalas eat eucalypts, cattle eat grasses, and Wedge-tailed Eagles eat sparrows. Most predation of plants can be attributed to invertebrate animals, which have been shown to remove from 10 to 40% of the total year's growth in eucalypts. Predation can dramatically reduce the range of a species from its potential. For example, in Tasmania the Spur Velliea (*Velliea paradoxa*) is largely confined to those places free of grazing by sheep and cattle.

Interference between species can be subtle. Some succulent species have the ability to accumulate and then excrete salt into the soil, thereby excluding less salt-tolerant competitors. Other species indulge in a form of chemical warfare known as ALLELOPATHY. Substances that are produced in their leaves, stems or roots suppress or kill other plant species. For example, many introduced grasses and herbs are susceptible to the chemicals washed from eucalypt leaves (leaf leachates). That is why it is so difficult to establish a lawn beneath a eucalypt tree, despite the fact that luxurious growth is usually found beneath similar eucalypts in the bush. Chemical defence is probably more frequent than chemical offense. The chemical complexities of the foliage of plants are thought to have largely evolved to defend them against predators.

Not all interactions between organisms involve a loss for one of the participants. In the northern Australian rainforest the trees provide the environment for a wide range of epiphytic species, such as orchids and bird nest ferns. Some plants in the rainforest have evolved tunnel systems in their stems or tubers which encourage the establishment of ant colonies. These ants reciprocate the favour by protecting the plant against herbivores. Similarly, old eucalypts often develop hollows in their stems and branches. The birds and mammals that nest in these hollows import food for their young, whose wastes fertilise the tree. Many fungi feed off the root systems of higher plants at the same time as converting phosphorus from a non-available to an available form. These mycorrhizal associations are vital in the nutrient economy of the bush. The licentious pseudocopulatory orchids give pleasure to individual invertebrates at the same time as ensuring their own pollination. Many Australian plants have elaeosomes, edible attachments to less edible seeds. Ants drag elaeosome and seed away from the parent plant. They benefit from the elaeosome; the plant species benefits from the dispersal.

The regeneration niche and vegetation dynamics

Individual plants are most vulnerable when they are germinants and small seedlings. Their success at these early stages of life can be improved by adaptations that restrict germination to the most favourable places and times. Plant species differ markedly in their germination requirements. Many have an inherent dormancy which is only released by a particular environmental signal or disturbance. For example, most wattle species have hard-coated seeds that will only germinate after heating or mechanical damage. This means that they germinate *en masse* after fire, when conditions are most propitious for survival. The seeds of some high altitude eucalypts will not germinate until they have been subject to a period of low temperatures in moist conditions. This ensures that they will germinate in spring. Many of the species in the southern heath family (Epacridaceae) produce their best germination in the regurgitates or droppings of birds. This maximises the chances of germination occurring in a location remote from the parent plant.

The regeneration niche of a plant species will be more spatially restricted than its fundamental niche (the climatic/edaphic envelope within which a species could survive and reproduce in the absence of competition). The regeneration niche consists of the set of conditions that allow it to germinate and survive to reproductive age. These include those aspects of the biophysical environment that are described above, but also include particular disturbance regimes.

Disturbance may be either ENDOGENOUS or EXOGENOUS. Endogenous disturbance is created by the growth, senescence and death of elements of the ecosystem. A good example is the mechanical collapse of a tree as a result of old age. The soil disturbance caused by native animals, such as bandicoots, that dig for food provides another example. Exogenous disturbance comes from outside the ecosystem. Fire, wind-throw, flood, drought, recreational bulldozing and killing frost are examples of such disturbance. As indicated by the word 'regime' above, it is not only the presence of suitable disturbance that is critical, but also its spatial and temporal patterning. An appropriate disturbance will go unused by a species that has completed a life cycle initiated by a previous such event, and is not within dispersal distance.

If exogenously disturbed vegetation is then only subject to endogenous disturbance, after directional vegetation change (succession), it will attain a long-term equilibrium, in which the demise of particular individuals of any species will be balanced by the establishment of other individuals. This equilibrium has been called the climax vegetation. This will only happen at particular time/space scales. For example, if the vegetation of one square metre of forest was followed through 200 years it might appear highly unstable, yet, if ten square kilometres were followed through the same time period, or the same area were followed through a millenium, the essential stability would be revealed. In climax forests much of the regeneration takes place through gap phase replacement, that is, new individuals establish themselves in gaps created by the demise of older individuals.

Most of the vegetation of Australia is not climax vegetation in the above sense, being dependent for its maintenance on a substantial component of exogenous disturbance. In the terms of Clementsian successional theory, it is disclimax vegetation, with the movement towards climax being constantly deflected by repeated exogenous disturbance. Much Australian vegetation is so well adapted to frequent exogenous disturbance, particularly by fire, that floristic change following disturbance is minimal, and only the relative abundances of species change. Such a pattern of change is described by the initial composition model of succession. It does not fit the classical relay floristics model, in which groups of species, apparently altruistically, hand on the baton to another group, this process continuing until the climax group attains the finishing line.

Relay floristics does occur in Australia in situations where new land is created by deposition, or soils are created by the colonisation of rock. Thus, in south-eastern Australia, the native sand binders, *Austrofestuca littoralis* and *Spinifex sericeus*, trap sand on beaches, forming dunes. The stabilisation of the sand disadvantages them and allows the invasion of salt-resistant shrub species, such as *Helichrysum paralium* and *Acacia sophorae*. These species, rainfall, and time modify the soil so that other shrubs, trees and herbs can invade and thereby displace their predecessors. However, the completion of all stages of successions such as this are the exception rather than the rule. Coasts and other places where deposition occurs are geomorphically highly dynamic. Just as a good succession gets going, the pattern of long shore drift changes and waves demolish the dune.

Relay floristics is not really an example of altruism. Ruderal species, which usually dominate the earliest stage of this successional pattern, have evolved to take advantage of disturbed or new ground. Their survival attests to the continued availability, within dispersal distances, of such environments. Each set of species in the race are in their particular stage because the environment in that stage provides their regeneration niche.

Endogenous vegetation change is not necessarily linear, inexorably moving towards the climax condition. Interactions between plants and their environment can create cyclic successional patterns. A good example of such patterns occurs on coastal sand dunes, where isolated plants trap sand causing the formation of a mound that is colonised by other species. The mound continues to grow until the shifts in wind patterns occasioned by the existence of other mounds lead to erosion and destabilisation of its base, and to its eventual loss.

Dispersal and migration

Plant species can only occupy their potential range if they can get to all parts of it at times when conditions are suitable for establishment. The part or whole of the plant that moves from one place to another in the act of dispersal is known as the PROPAGULE or disseminule. Most propagules achieve little distance from their female parent. For example, almost all eucalypt seed falls within a distance twice the height of the maternal parent. However, it is the exceptional circumstances, rather than the normal, that tend to be important in dispersal and migration. Eucalypt seed may be carried much longer distances than twice tree height in the fierce convectional updrafts of a raging fire, or along a stream. It even seems likely that eucalypts can migrate through hybridisation and then stabilising selection.

Other Australian species are better adapted to long-distance dispersal than eucalypts. Many attach themselves to animals and birds (ectozoic dispersal), or are carried viable in their digestive systems (endozoic dispersal). A large number of species have seed so small or aerodynamically well-designed that it is carried long distances in the breeze. Others, like Huon Pine (*Lagarostrobus franklinii*), have seeds that remain bouyant and viable in water.

Despite the many adaptations for long distance dispersal and the many rare events that could disperse unadapted disseminules over large distances, relatively diminutive barriers can block dispersal of most plant species for very long time periods. For example, the calcareous Nullabor Plain has been effective in isolating the flora of the siliceous soils of the south-west of Western Australia from the flora of siliceous country in south-eastern Australia, to the extent that they have few higher plant species in common.

The number and variety of disseminules in motion decline exponentially with distance from a source region. Thus, oceanic islands inevitably have poor and unbalanced floras compared to islands close off shore. The theory of island biogeography suggests that each island has an equilibrium species number which is a function of a migration rate, controlled by distance from source region, and an area-controlled extinction rate. If more species migrate the resources of the island are shared more widely, making the rarer species more liable to extinction in the normal course of population fluctuations. If more species become extinct, the chances of successful migration increase.

ECOSYSTEMS AND COMMUNITIES

An ecosystem is an ecological system. It can be at any scale: a drop of water, a pond, a forest, a saltmarsh or the biosphere (that part of the earth and its atmospheric envelope that supports life). An ecosystem contains both biotic and abiotic components which interact in various flows of energy and materials. The network of eaters and eaten called the food web is a prime example of this sort of systematic interaction. Ecosystems have boundaries with other ecosystems across which energy and materials exchanges take place. Most natural ecosystems are so complex that we can never hope to completely understand the nature of their energy and material flows. However, we can hope to learn something of the consequences of ecosystem changes by focusing on what appear to be critical interactions. For example, we now have more than a rudimentary knowledge of the consequences of the forest logging cycle on the stores and flows of the critical macronutrients phosphorus and nitrogen.

To understand the nature of ecological systems we need to be aware of feedback and chaotic effects. Negative feedback occurs when the system reaction to a change reverses that change. For example, the removal of nutrient ions from part of the soil solution steepens the concentration gradient and therefore hastens the movement of more nutrient ions from weathering minerals. Systems with negative feedback mechanisms are highly stable, comparable to a pea in the bottom of a bowl. Positive feedback results when a system reaction to a change reinforces that change. For example, the passage of a fire through a rainforest opens up the canopy, allowing greater drying of fuel than was previously the case, and thereby increasing the chances of the occurrence of a later fire. Systems with positive feedback mechanisms are in unstable equilibrium, equivalent to a pea perched on the top of an upturned bowl. Chaotic effects happen at particular space/time scales where positive feedback is so strong that unmeasurable initial changes can shift a system into dramatically different pathways.

While the term 'ecosystem' incorporates the inorganic and organic and provides a focus on flows, the term 'community' simply relates to living organisms that repeatedly cohabit. Thus, a plant community would consist of a particular assemblage of plants or types of plants that recurred in a reasonably consistent manner in different parts of the landscape. Plant communities can be defined on the basis of dominance of a particular life form, their geometry (structure), species dominance, the number and characteristics of the vegetation strata, and the repeating coincidence of species (phytosociology). For example, a forest is a plant community dominated by trees with crowns wider than the spaces between crowns. Forest may be subdivided on the basis of its height and canopy coverage. The most commonly used structural scheme in Australia is that of Specht. This uses the projective foliage cover (area of shadow with sun directly overhead) classes of 0–10%, 10–30%, 30–70% and 70–100% and the height classes of 0–2 m, 2–8 m, 8–30 m and 30 m +. This has classes titled tall closed-forest (>30 m tall + >70% canopy cover), closed-forest (8–30 m tall + >70% canopy cover), tall open-forest (>30 m tall + 30–70% canopy cover) and open-forest (8–30 m tall + 30–70% canopy cover). Tall open-forest has been divided according to the nature of the understorey (e.g. grassy tall open-forest), the dominant species (e.g. *Eucalyptus globulus* grassy tall open-forest) and phytosociology (e.g. *Eucalyptus globulus – Poa labillardieri – Veronica gracilis* grassy tall open-forest).

While there are many positive and negative interactions between species (see biotic factor above) that promote a tendency towards consistent association between plant species of different life form, communities typically intergrade with one another. This means that, with a few exceptions, communities are essentially arbitrary sections of multi-dimensional continua. The continuous variation exhibited by communities contrasts with the discontinuities that characterise species boundaries. However, arbitrariness in definition does not necessarily translate into uselessness. Communities may intergrade with others, but be easily defined and good predictors and descriptors. Also, discontinuities in environmental conditions, or discontinuities created by the edges of the ranges of dominant species, often create sharp boundaries in vegetation, such as those encountered at the edge of wetlands, at the climatic tree line and on geological boundaries. These can create relatively well-defined units for mapping and description.

RICHNESS AND DIVERSITY

One of the great questions of ecology is 'how do so many species manage to coexist?'. A corollary question is, 'why are there more species in one spot than another?'.

The number of species found in a defined local area is known as its SPECIES RICHNESS or alpha diversity. Species richness varies dramatically. For example, some south-eastern Australian grassy woodlands have more than 80 higher plant species in an area of ten square metres, while some saltmarshes in the same region have only one species in the same area.

Viewed globally, terrestrial species richness is greatest in the tropical rainforest. Areas with plant species richnesses that rival the tropical rainforest are found in the southern temperate Mediterranean heathlands, particularly the fynbosch of South Africa and the kwongan of Western Australia. On a global scale we would expect that the areas of highest species richness would have avoided the mass extinction episodes associated with the Quaternary climatic fluctuations and also would have had the longest periods in which speciation could take place. The patterns conform to this supposition with the European flora, pushed into the Mediterranean Sea during glacial times, being impoverished compared to the East Asian flora, which simply migrated to Vietnam. The tropics suffered least from climatic fluctuations, while receiving enough environmental stimulus to encourage allopatric speciation.

At a more local scale, species richness tends to be greatest on sites of intermediate disturbance and resource availability, because relatively few species are adapted to the extremes of resource availability or disturbance, no matter what the range of such conditions is in any area.

BETA DIVERSITY is a measure of the turnover of species along an environmental gradient. It is measured in half changes, that is, the distance along the gradient at which half the original species are replaced by new species. So, at exactly two half changes there would be no species in common.

Both alpha and beta diversity are distinct from the concept of diversity as a tendency towards polydominance. In this sense a highly diverse community would be one in which there were many species of equal abundance, whereas a non-diverse community could have the same number of species, but one species would constitute most of the individuals or biomass. Confusing terminology is a special feature of ecology.

AN INTEGRATED EXPLANATION OF SPECIES DISTRIBUTION PATTERNS

Figure 2.1 consists of a series of circles symbolising the progressive restriction of species from their fundamental niche by the factors

Figure 2.1 A model of the restriction of the distribution of species from their fundamental niche.

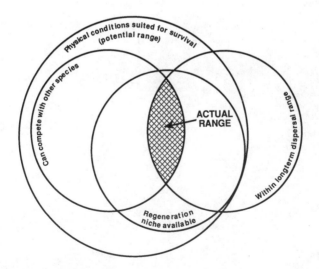

discussed above. A plant species might have sufficient moisture, heat, light and mineral nutrition anywhere in Australia. However, it will not occupy its fundamental niche if other species prey upon, or otherwise interfere with it, to the extent that it cannot reproduce. It could be further reduced in area by the lack of the particular disturbance regimes that are necessary for its regeneration, or the presence of disturbance regimes that lead to its demise. Within the remaining area still theoretically available to this notional species, parts might not be accessible to its disseminules because of their lack of long-distance dispersal ability and the distances involved.

A CONCLUDING PREAMBLE

The ecological patterns created by the natural processes described and discussed above can and have been dramatically affected by human beings. While our species is obviously as much a product of evolutionary and ecological processes as any other, we have gained evolutionary responsibility, or irresponsibility, through our ability to modify the very nature of the planet. Evolutionary and ecological processes will limit this modification through feedback effects on our species. In the interim we would be wise to attempt to understand our impacts and their implications.

■

3

Glaciers and Aborigines

PREHUMAN ENVIRONMENTAL HISTORY

Australia was once part of the supercontinent of Gondwana, a giant land mass that also included Antarctica, South America, Africa and India. About 45 million years ago the Australian plate drifted northwards, a process that resulted in the creation of the circumpolar ocean current, which in turn caused the icing over of Antarctica. At the time of separation from Antarctica most of the Australian continent appeared to have been forested, despite the high latitude and long nights. This forest cover seems to have declined somewhat as the result of climatic changes that took place in the middle years of the Miocene (24 million to 5.2 million years before present (BP)), and was massively disrupted by the climatic fluctuations of the last two million odd years.

During the two million years of the Quaternary, cold and dryness has descended every 80,000 years or so, and has regularly continued for many tens of thousands of years before warmth and wetness ascended again. During the relatively brief warmer periods forests covered only about 12% of the continent. During the coldest periods forests almost disappeared. The last ten thousand years of the Quaternary, known as the Holocene, has been one of the relatively warm and wet periods.

In the south-eastern corner of the continent, in particular, there must have been considerable biological chaos in the first of the Glacial periods, with new species evolving to take advantage of colder and drier conditions and with others retreating north or to lower areas to gain warmth, some perhaps becoming extinct because of lack of mobility or lack of pathways. Although it might be suspected that most of the speciation and extinction resulting from massive cyclical changes in temperature, precipitation and land area would have occurred in the early Quaternary, there is evidence of much later plant and animal extinctions. Many of these are known to have occurred after human beings invaded Australia.

SOURCES OF EVIDENCE ON
THE PREHISTORIC HUMAN IMPACT

There are many different sources of information on prehistoric environments and the human impact. Records of Quaternary climatic oscillations have been obtained from cores in marine and terrestrial sediment, from coral reefs and caves and from plant macrofossils and pollen. In all cases a workable version of the uniformitarian assumption has been adopted, this being that many physical relationships and characteristics of biological entities in the past can be deduced from their present condition, that is, that species and communities have not changed their ecological responses. While most of the physical records of climatic change are largely independent of human activity, the vegetation that produced the macro and microfossils was highly sensitive to human activity, especially fire. Thus, it is theoretically possible to separate the effects of climatic change on vegetation from the effects of human beings by using data from the two sets of records. Human influence may be deduced where the more recent climatic cycle/s result in the expression of a different plant fossil record from earlier identical cycles. However, this is a very rare conjunction. Most influences are inferred from historical or ecological analogies.

The influence of humans may be further deduced from increases in the charcoal content of the deposit, although increased charcoal could reflect either the result of vegetation change or its cause. Interpreting changes in the abundance of charcoal can also be made difficult by the fact that little charcoal is produced when fires burn so frequently that shrubs and trees are largely replaced by herbs, and because well-humified organic matter bears a close resemblance to charcoal.

Interpreting macrofossil or pollen sequences in terms of vegetation also entails problems. Macrofossils from different species tend to accumulate differentially in different parts of waterbodies. Thus, the macrofossils of one species may be prominent in recent sediments, and those of another in older sediments, because of a change in the size of the waterbody not a change in the surrounding vegetation.

Pollen profiles can be difficult to interpret because pollen is produced in different amounts by different plant species; because the pollen of some species travels much further than that of other species; because the pollen of large groups of related species is virtually identical; and because the pollen of some species preserves better and longer than that of others. The pollen analyst tries to relate the proportions of the various types of pollen that are present at a particular depth to the vegetation from which they were derived by comparing the fossil pollen rain with that found today in a range of vegetation types. However, the contemporary pollen rain can vary enormously over very short distances, as most pollen falls close to its source. There is thus potential for considerable disagreement on the meaning of a pollen profile, which is related to lack of detailed knowledge of the contemporary pollen rain. Our poor knowledge of the contemporary physiological and ecological responses of species to environmental gradients does not help in the interpretation process.

It is difficult to draw structural and dominance conclusions from the pollen record because, with a single core, a mosaic of taxa cannot be separated from a community of taxa, and because most of the more frequent pollen taxa adopt a diversity of lifeforms. For example, both *Nothofagus* and *Eucalyptus* can dominate vegetation as a shrub or a tree, and the former can be a subordinate species in both lifeforms. Composites can be trees, shrubs or herbs, and dominant or subordinate in all lifeforms.

Despite these problems, in an exciting surge of recent research, pollen analysts have virtually created our understanding of the vegetation of the Holocene and late Pleistocene, and their conclusions on the broad nature of vegetation change during this period have gained wide scientific acceptance.

The current vegetation can also provide major insights into the impact of Aborigines on past vegetation. The vegetation in many areas of Australia is still dominated by plants that were growing before the European invasion. These plants contain records of past fire regimes, in the form of charcoal on their growth rings. Their ecological requirements give us clues to the nature of prehistoric land management. Vegetation changes that are in process today as the result of the elimination of Aboriginal land-use practices can also give us clues to the past manner of management, but only if we fully understand the European disturbance regimes and the nature of the ecological responses to them.

While some knowledge of the results of prehistoric human activity on vegetation can be gained from pollen and macrofossil records and the current vegetation, our understanding of the behaviours that produced these environmental changes relies on the archaeological record and extrapolation from historical and ethnographic observations.

Given the right set of depositional conditions, human artefacts and waste can accumulate on a site as pollen does. Largely confined to those made from rock or bone, the surviving tools can be used to deduce major interactions with the environment. For example, the tool kit associated with the exploitation of scale fish differs from that associated with the exploitation of shellfish. The nature of the animal component of diet can be at least partly deduced from bones, carapaces and shells. Unfortunately, there are usually few or no traces of vegetable foodstuffs in either archaeological deposits or art work.

The observations of Aboriginal behaviour made by European explorers and early colonists suffered from cultural prejudice. Except for the earliest maritime explorers, these observers, and the latter day ethnographers, were viewing societies already disrupted by a massive wave of disease, and, in later times, by murder and cultural suppression. Also, the Aboriginal culture was neither monolithic nor static — for example, the Tasmanian Aborigines adopted the domestic dog almost immediately and the culture of the northern Australian Aborigines was influenced by the Macassan traders. Thus, the historical and ethnographic record must be interpreted with caution when extrapolating to prehistoric times.

THE ARRIVAL OF *HOMO SAPIENS*

One of the most vexed questions in modern Australian archaeology is the date of arrival of humans in Australia. There is no doubt that people have occupied the continent for at least 40,000 years, the limit of dating based on the radioactive decay of carbon-14. At this time they used fire to produce the carbon that, in turn, produced the date.

Forty thousand years ago Tasmania was a peninsula of mainland Australia, New Guinea a northern appendage and Asia much closer than it is today. However, all these statements hold true for most of the last two million years. Even with present high sea levels the opportunity for migration would not be absent. It was certainly present for most of the millennia before 40,000 BP during which human beings were found on the planet.

Rhys Jones and his co-workers believe that they have extended occupation back to 60,000 BP using the thermoluminescence method of dating on a northern Australian archaeological deposit, although many of their colleagues are reluctant to concede the veracity of this date.

Indirect evidence, from a discordance between climatic cycles and vegetation that is found in pollen and charcoal diagrams from Lake George near present-day Canberra and from the present-day seabed off northeastern Queensland, has been taken to suggest that human occupance of Australia occurred more than 120,000 years ago.

Gurdip Singh has described and interpreted some of the major changes in the composition of the charcoal, pollen and spore rain that have occurred while eight and a half metres of sediments have been deposited on the floor of Lake George. This pollen record covers three and a bit glacials and three and a bit interglacials, the bit of an interglacial being the last 10,000 years. After 40,000 BP carbon-14 dating gives good evidence of the age of the pollen profile. The older dates assume that the record does in fact show the long-term sequence of glacial and interglacial, not shorter term fluctuations. These dates have been obtained by analogy from those obtained by other means from deep-sea cores. The tentatively advanced hypothesis was that human beings arrived during the glacial before last (i.e. more than 128,000 years ago). Their landscape management practices, largely involving the use of fire, led to a dramatic decline of *Allocasuarina* and a dramatic increase in grass and eucalypt pollen in the following two interglacials.

In 1992 Peter Kershaw and his co-workers suggested that dramatic changes in the proportions of pollen types, and their correlation with increased charcoal levels, in a deep-sea core off north-east Queensland, might support the above interpretation of the Lake George record. The ratio of oxygen isotopes in these ocean deposits was be used to gain a more reliable date for this change than could be gained from Lake George. The change occurred approximately 150,000 years ago.

None of the three millennia proposed for the arrival of human beings in Australia had climatic conditions or sea levels similar to those of today. Aborigines have modified the vegetation of Australia during conditions close to the climatic extremes of the Quaternary; the height of the Last Glacial approximately 18,000 years ago, and the early Holocene, approximately 8000 years ago. The period between 18,000 and 8000 year ago is thought to have seen an increase of approximately 6°C in mean annual temperatures in southern Australian and a doubling of rainfall. The rate of change would have led to almost imperceptible changes in environment during the lifetime of any one generation. However, the rise of the sea level to its present height 6500 years ago, from its nadir of 120 m below the present level at the height of the glacial, was passed on in stories of the dreamtime.

There is no doubt that the invasion of Australia by humans engaged in gathering and hunting had a profound effect on the ecosystems of the continent. There is doubt about the exact nature of the impact. It is unlikely that this doubt will be resolved in the near future given the paucity of evidence available from 40,000–150,000 years ago.

THE NATURE OF THE PREHISTORIC HUMAN IMPACT

Edmund Curr, in *Recollections of Squatting in Victoria*, his book on the pastoralist invasion of the western basalt plains, expressed the opinion that few people would have changed their landscape as much as the Aborigines of Australia, an opinion that has gained enormous ecological, archaeological and anthropological support in recent decades. The Aborigines used fire as a tool to work their landscape. Given enough time, fire can be as successful as a bulldozer in changing a forest to a grassland. Firing promotes the growth of plants with edible parts, is useful for attracting, herding and killing game, and can used for cooking, to provide warmth, repel bloodsucking insects and clear paths through thick or thorny scrub.

There is no doubt that Aborigines set fire to the bush purposefully and frequently, and with a high degree of control. There are several early, and many contemporary, accounts of Aborigines using fire to protect particular features, such as copses useful for sheltering game and forests rich in particular food resources or containing sacred sites.

The skilful use of fire as a tool for land management did not preclude what today we would consider the occasional ecological disaster. For example, there is good evidence that burning caused the loss of communities dominated by the fire-susceptible Deciduous Beech (*Nothofagus gunnii*) in western Tasmania 500 years ago. This fire may have been deliberately lit to clear an easy passage through seldom-visited country, it may have been an escape from a fire burnt for other purposes, or it may have been lit by lightning.

The exact extent to which Aboriginal burning eliminated, created or shifted the boundaries between vegetation types will probably always remain a matter for supposition. However, there is no need for supposition in some particular instances. All along the eastern seaboard of Australia and in Tasmania, open grassy areas in a matrix of rainforest are being invaded by rainforest species. There seems little doubt that these grasslands largely resulted from Aboriginal burning, especially in Tasmania where lightning-lit fire is a rarity. In northern Australia the cessation or reduction of the early dry season patch burning practised by the gatherers and hunters has resulted in extensive late dry season fires, which are imperilling stands of native pine (*Callitris* spp.).

The role of gathering in modifying the vegetation has tended to be underestimated. Yet more than 140 plant species were used by Aborigines in Arnhem Land, and at the other extreme of the continent, in the grasslands and grassy woodlands of Tasmania, there are more than 100 species that are known to have been used by Aborigines somewhere in Australia. The utilisation of a large number of these species involved substantial environmental disturbance. In the kwongan heathlands in what is now the wheat belt of Western Australia, observations were made of gathering activities that involved digging holes more than a metre deep. Beth Gott argues that the Aboriginal people of the western basalt plains of Victoria engaged in activities closely related of farming. In their utilisation of the once abundant Myniong (*Microseris lanceolata*), they not only dug to remove the tubers but also replaced the tops that were capable of forming the next years harvest. The first European explorers reported lines of women and children collecting yams and noted the softness of the soil, possibly the result of continous digging. Massive amounts of tubers and corms were obtained from wetlands, and, in the forest, the rhizomes of Bracken (*Pteridium esculentum*) were used as a source of carbohydrate. Acacia seeds were often harvested and cooked in one operation by burning the plant. This would, of course, stimulate the germination of seeds held in the soil store from previous years. Other seeds and fruits were widely harvested. For example, Aborigines collected in great social gatherings in south Queensland to harvest the seeds of the Bunya Pine (*Araucaria bidwillii*).

In particular instances, the gathering activities of Aborigines may have led to the evolution of particular food species. The two species of kangaroo apple (*Solanum aviculare* and *S. vescum*) make up one of the few native Australian plants that has found a use in contemporary agriculture. It is a weakly woody and short-lived shrub with large dissected leaves which resemble those of cassava and marijuana. Its purple flowers, if appropriately pollinated, become hard green fruits which soften into yellow and orange. In the early stages of development the fruits are poisonous, like most green fruits of other members of the potato family. When ripe they are passable, with a taste combining the apple and the cooked potato. However, its agricultural use is not for the table, but rather to provide steroids for the birth control needs of the citizens of Russia.

The kangaroo apple is one of the many plants that exchanges food for seed transport. Its ripe seeds will pass through the digestive and excretory systems of animals and birds, to emerge in its own fertiliser with enhanced viability. Thus the advantage of poisonous unripe fruits is to postpone transport elsewhere until such transport can be exploited for dispersal. Not being long-lived, and growing easily only on fertile or recently-burned ground, the kangaroo apple depends on attracting dispersal vehicles.

The Aborigines played a major part in this transport, and in doing so must have selected for the fruits most palatable for human beings and for the seeds most able to establish on the types of disturbed ground found around Aboriginal encampments. Thus, the kangaroo apple was an unconscious domesticate, awaiting agricultural people, and being dependent for its abundance on the activities of human beings. Today, the species is found where suburbia merges with bush, in the highly fertile ground of bird rookeries and, as a weed, on the fertile hopfield soils. Thus, it is still largely a human dependent, although its fruits lack the allure of Eurasian and American domesticates.

Soil disturbance, like that created during the gathering of bulbs and tubers, is a critical element in the regeneration niche of many of today's most rare and threatened species. The absence of ground-disturbing gathering activity may, along with the extinction or depletion of many digging marsupials, account for this current rarity.

Aborigines may have caused indirect changes to the vegetation of Australia through their impact on animal populations. Like the Americas and Eurasia, Australia experienced a massive episode of extinction of the large animals known as the megafauna. A large number of species of Australian megafauna survived many thousands of years past first human occupance of the continent, and were recorded in Aboriginal cave paintings. Robert Ardrey has argued in his hunting hypothesis that the survival of the megafauna in Africa and its extinction elsewhere reflects a coevolution in Africa, the home of humanity, compared to a sudden high technology onslaught elsewhere. It seems difficult to argue that the extinction of the Australian megafauna was a phenomenon totally unrelated to human activity. The megafauna survived through a large number of climatic cycles of similar intensity to the last glacial/interglacial cycle, the only difference in the last being the presence of human beings. The mechanism of human-induced extinction was obviously not just hunting, as the extinctions took place over many tens of thousands of years. Modification of the environment and hunting probably combined with climatic deterioration to prevent survival.

Two almost certain cases of extinction due to human activities related to the introduction of the Dingo to mainland Australia 4000 years ago. This was rapidly followed by the loss of both the Thylacine and the Tasmanian Devil, which only survived in Tasmania, which by that stage had become an island.

It is difficult to deduce the impact of the megafauna extinctions on the vegetation of Australia. Some ecologists have suggested that large-fruited plants, such as *Pandanus* and *Macrozamia*, may have lost their main dispersal agents and therefore become less able to expand their ranges in response to climatic amelioration. However, more subtle changes undoubtedly occurred, including selective shifts in grazing pressure and alterations in soil disturbance regimes. The interconnectedness of the components of the ecosystem would necessarily result in some major change. I illustrate such changes from a contemporary example of local extinction studied by David Ashton and his students in the Botany Department at the University of Melbourne.

Sherbrooke Forest is an area of century-old *Eucalyptus regnans* on the south-facing slopes of the Dandenong Ranges. Both before and since the range was engulfed by suburbia, Sherbrooke has been a favoured venue for the mountain picnic and ramble so beloved by many city dwellers.

Their strolls among ferns and beneath giant straight-boled eucalypts have been considerably enhanced by the populations of the Superb Lyrebird that worked the litter-covered ground. As well as providing excellent imitations of chain saws and cars changing gear, the lyrebirds perform attractive mating ceremonies that can be viewed by the very quiet stoller. Sherbrooke also used to be the home for large populations of wombats and wallabies. The lyrebird survived the cats, dogs and motor vehicles that came with the encroachment of Melbourne. The wombats and wallabies were severely reduced in number, becoming almost extinct in the west of the forest.

One of the favourite foods of the Sherbrooke wombats was Wire Grass (*Tetrarrhena juncea*), an aptly named species that has the potential to form tangled masses more than one metre tall. It promptly fulfilled its potential following the demise of the local wombats, and ground that was previously ideal for the scratching of the lyrebirds became inaccessible. Their habitat having shrunk drastically, the lyrebirds have suffered a population crash.

The impacts on wombats and lyrebirds undoubtedly have further implications. For example, the soil fauna and flora associated with a Wire Grass understorey might be very different from those associated with an understorey of leaf litter constantly worked by lyrebirds. Thus, the rate at which fallen litter is broken down into a form available for consumption by the live plants might have changed, and this decrease could affect the growth rates of the trees of the forest. Any change will reverberate through all the organisms whose existence is at least partly dependent on those organisms at first, second, third to nth remove in the food web from the organism with the initial population change. Eventually all these changes will become insignificant, but when they do so the biological and physical world could be very different, and in highly unexpected ways, from the ecosystem within which the initial change took place. All of which serves to illustrate the point that the Aborigines must have drastically changed Australian ecosystems by changing the patterns of predation on animal populations.

People in harmony with nature?

There is a popular view that Aborigines were in harmony with their environment; in an idyllic balance between people and nature, reinforced by culture. The belief systems of the hunters and gatherers were undoubtedly directed towards the maintenance of their resources. The land, in its broadest sense, seems to have been regarded as inseparable from the people. The archaeological record shows that people adapted to the full range of Australian environments and maintained flows of resources for millennia. It also reveals some substantial shifts in technology and population, as with the shift from utilisation of the fin fish resource in Tasmania documented by Rhys Jones.

Belief structures pass through a similar selective screen to genotypes. For example, a society that believed that the total level of resource utilisation could increase indefinitely in a finite world would be doomed to extinction, as would a society whose belief structure prevented procreation. Beliefs are rarely based on logical analyses of ecological situations. Rather, those beliefs that allow the continuation of a society survive and are internalised in its members, while the societies (although not necessarily genotypes) with maladaptive beliefs disappear. Any change in technology or environment may make some elements of a belief structure, or the whole belief structure, maladaptive.

Aboriginal people in Australia attained an equilibrium with their environment after their activities caused massive changes in the biotic environment, and this balance relied upon a constant type and intensity of cultural interaction with the landscape. The environment may have adjusted more to their practices than their practices to the environment. Aboriginal societies were and are prepared to adopt new technology where such technology is consistent with their belief structures. The likelihood that these belief structures have evolved to allow a constant relationship with the rest of the ecosystem cannot be taken to prove that any culturally consistent innovations would maintain this harmony. For example, the Dingo, and later the dog, were readily incorporated in the Aboriginal socioeconomy. Their impact on the native biota has been far from insignificant. If the gatherers and hunters had known of the inevitability of this impact, would they have hunted the Dingo to extinction before it occupied the continent? Perhaps not, as the value of the Dingo to their socioeconomy almost certainly outweighed that of the Tasmanian Devil and Thylacine.

Current Aboriginal interaction with the environment

Aboriginal people currently have ownership of a large part of the remaining natural vegetation of Australia. Their culture and belief systems are still dramatically different from those of the European invaders and the more recent immigrants, yet have substantially changed from those of the late eighteenth century. Over large segments of the country the traditional knowledge of the Aboriginal people has been lost through genocide and cultural disruption. However, in the north and the desert, where most Aboriginal land is situated, much traditional knowledge survives.

The survival of knowledge does not necessarily translate into a maintenance of pre-invasion management regimes. Populations of Aborigines are both depleted and concentrated, and elements of modern technology have replaced most of the original tool kit. Recently introduced animals that are known to deleteriously affect native species have become major food resources, creating resistance to programs directed towards their control or elimination. Thus, the elimination of buffalo from the Top End was resisted by the local Aboriginal people, and in central Australia the control of rabbit populations is similarly regarded negatively. Poverty and European pressure can conspire to force decisions that resemble those of developers. The Tiwi people of Bathurst Island are engaged in converting native forest and grassland into introduced pine plantations and seem likely to permit extensive sand mining operations in their coastal sand dune systems.

We have no reason to believe that Aboriginal people will be, on average, any better custodians of the natural values of the land than any other owners. Their new tools and economies point towards a future equilibrium that might lack much of the biological diversity that survives today. Nevertheless, Aboriginal knowledge is vital in properly managing biological diversity, and the relative greenness of Australia in environmental attitudes is some indication that the Aboriginal feeling of oneness with the land has permeated into the general consciousness.

Another Concluding Preamble

The enormous changes to the bush that have been briefly described above were relatively minor in speed, extent and intensity compared to the changes that have occurred in the last two hundred years. Enormous areas of bush of particular types have been replaced by pastures, agricultural crops, dams, cities, roads and railways. Firing patterns have changed from those of the Aborigine, creating different types of bush from those found two hundred years ago. Much of the forest not replaced by farm or Radiata Pine has been modified by slash and burn silviculturalists. Plagues of root-rotting fungi, hard-hooved animals and opportunistic plants are in the process of transmogrifying most of the remaining bush. Even the very popularity of the bush for walkers and nature lovers is creating localised disturbance.

It has become necessary to design reserves for the perpetuation of the bush and try to manage species and plant communities to aid their survival. The prognosis for survival varies. The rest of this book is an attempt to describe and understand this variation.

■

4

Bush destruction and the creation of cultural vegetation

Seminatural and natural bush still covers most of Australia (Figure 4.1). Nevertheless, over large areas it is possible to sit on a hill and observe few or no individuals of any plant species that was present in the area before European settlement. The transformation from bush to a vegetation that directs the energies of the sun to the multifarious purposes of the human race has engaged much of the physical and intellectual energies of the European Australians. The successes of the pioneering drive against the bush were celebrated in the lavish construction of cities with gardens derivative of Europe and bereft of reminders of an alien land. The natural grandeur of Sydney Harbour has been repeatedly and proudly mimicked as bush has met a watery grave in order to service the thirsty and space-prodigal Australian city. Bush long resistant to conversion to the ends of humans has been recently defeated by science and the agricultural developer. However, agricultural resources have often been less than perceived by the developers, who have failed and left their land as neither bush nor farm.

THE INCIDENCE AND MAGNITUDE OF EUROPEAN BUSH DESTRUCTION

Agricultural land clearance has affected approximately 6.1% of the land area of Australia, with another 1.4% devoted to roads, rail, impoundments

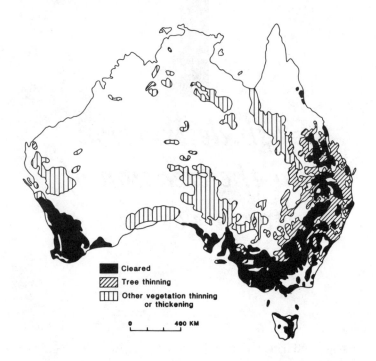

Figure 4.1 The distribution of cleared and degraded natural vegetation in Australia (based on the maps in AUSLIG (1990). Atlas of Australian Resources: Volume 6 – Vegetation. AGPS, Canberra, and observations of the author).

cities and mines. The acts of land clearance, inundation and reclamation are characterised by the almost complete loss of the native macrophytes, although a subset of native invertebrates and the microflora may persist in water and soil.

The term 'clearance' encapsulates the attitude that sees the bush as an obstacle. After all the visible native vegetation has been removed, it is replaced by domesticates and introduced ruderals. Inundation, the euphemism for which is water conservation, converts terrestrial into aquatic ecosystems. However, its major impact on nature has been through the modification of the hydrological systems on which much wetland vegetation depends. The dirt apron found around the typical impoundment contrasts markedly with the profuse growth found on the margins of most natural freshwater wetlands. The artifical pattern of water rise and fall in impoundments usually exceeds the tolerance limits

Plate 4.1 A typical Australian rural landscape: improved pasture in the foreground, trees killed by inundation in the middleground and native vegetation surviving on the hills.

of the native wetland plants. Changes in dowstream flow patterns may also severely affect native vegetation. For example, the mitigation of flooding on the Murray River has led to tree invasion of grasslands previously maintained by flooding. Reclamation converts wetland into terrestrial environments. In this case, even the soil is artificial. Nothing is reclaimed in the normal sense of the word.

The 7.5% of the continent directly affected by clearance, inundation and reclamation represents one extreme of the continuum of European modification of the continent. At the other extreme, approximately 30% of the vegetation of the country remains closely similar in its species composition and physiognomy to that found by the first European invaders. The intermediate areas have been affected by changed disturbance regimes, introduced animals and plants and the downstream consequences of land clearance, with 2% of the continent affected by logging and 61% affected by stock grazing. The effects on the native plant species of long-term land degradation may ultimately be indistinguishable from the effects of clearance and cultivation.

The incidence of destruction of native vegetation has varied over the last two hundred years. The early years of settlement targeted the temperate grasslands and grassy woodlands (see below). The latter decades of the nineteenth century saw the loss of over one quarter of the 1788 forest cover of New South Wales, including the almost total demise of the Big Scrub, one of the largest stands of rainforest in Australia. In Victoria, most of the forest was cleared from the South Gippsland Hills. This rush of forest clearance has continued to the present. Forty-three

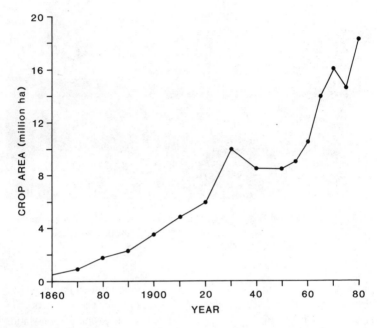

Figure 4.2 Changes in the area of crops and improved pasture in Australia.

percent of the forests present at first European settlement have been cleared for agriculture. The period from 1890 to the present has seen the Mallee, heath, Brigalow and Poplar Box encounter the exponential curve of clearance, at the same time as an increase in the area of agricultural land from 2.2 to over 9.0 million ha (Figure 4.2).

Bush resistance to clearance

The Australian bush has not been lacking in challenge to those who would break its soil for pasture and crop. Most land that received enough rainfall to make breaking the soil surface worthwhile was covered by tree and shrub species whose regenerative powers were tempered by frequent fire. The eucalypts are not easily eliminated. They will resprout from stumps or their large woody underground lignotubers if their aerial parts are destroyed. Removal of the stumps might mitigate this problem, but most forests and woodlands abound with small suppressed trees that are only too happy to replace their deceased relatives. Seedlings also vie for this honour. Eucalypt seed is held in woody fruits and is released in profusion when a branch or tree is defoliated by any cause, so seedlings densely occupy the disturbed ground of newly cleared land.

The elimination of eucalypt seedling and resprout does not end the problems. Many of the smaller shrubs and trees store hard-coated seed in the soil. The heat of a fire or the mechanical disturbance of a plough are equally effective in releasing these seeds from their suspended animation. The plough can also be effective in encouraging the proliferation of the many plant species capable of forming a new individual from a small portion of their root systems. Eliminating the native bracken (*Pteridium esculentum*) from a field by digging up the soil has the same notable lack of success that greets the same strategy when attempted with rhizomatous grasses in the vegetable garden.

Once exotic plants are established the problems do not cease. The Australian bush seldom develops the lushness of an English field. The most common plants are usually heavily protected by thorns, thick and chewy leaf surfaces or by an aromatic range of chemicals partly developed to repel the native herbivore. The native herbivore has had the evolutionary time to adapt to these hardships, but many of the larger species, and some of the less specialised of the small, exhibit considerable appetite when presented with a soft and green European crop. This enthusiasm for the marsupial equivalent of fast food still necessitates protective covers over vegetable gardens anywhere isolated in the bush, and was a source of considerable distress to the first settlers of Australia who were often forced to devour the herbivore rather than the crop.

THE TRANSFORMATION OF THE PLAINS

The first Australian farmers soon developed a good eye for the land where trees were thin, the native grasses lush in season and the soil black with organic matter, crumblike in texture and rich in nutrients. The word 'plain' was used to differentiate such grassy areas from the less agriculturally attractive types of bush. Some such plains were often situated on the steepest of slopes, although most occupied flat to undulating land. Thus, the native temperate grasslands and grassy woodlands were the first vegetation types to experience the agricultural transformation.

There are still parts of south-eastern Australia where scattered trees sit among the tough native tussocks of Snow Grass (*Poa* spp.), giving the impression of the landscape that existed when the first pastoralists and their flocks occupied these plains in the nineteenth century. However, the trees and Snow Grass are most probably doomed, and little else in the vegetation has survived over a century of sheep and cattle grazing.

The plains were transformed more insidiously than other vegetation types. They were and still are used for crop-growing, but much of their area has only been used for grazing stock. Most of the plains were dominated by the red-tinged leaves of the deep-rooted perennial Kangaroo Grass (*Themeda triandra*). The Kangaroo Grass was found in mixture with numerous other grass species, including Wallaby Grass (*Danthonia* spp.), Spear Grass (*Stipa* spp.) and Snow Grass, and a profusion of showy herbaceous daisies, peas and other forbs. Most of the plains appear to have been maintained as such by burning of sufficient frequency and intensity to make the probabilities of tree establishment extremely low. In some cases this could mean virtually no fire, because tree and shrub species have great difficulty in establishing in competition with a sward of grasses. Some of the plains may have been created by the occasional severe frosts, like that of 1857 which created a sea of dead eucalypts in the valleys of inland Tasmania. Other plains were certainly caused by firing; the remains of dead rainforest trees are found in amongst native grasses in many parts of Australia.

The savannah trees that gave so much character to the plains country have been dying out, a process glorified with the title of 'rural dieback'. There has been much research on rural dieback, which has been variously ascribed to the effects of drought, fertilisation, insect plagues and exposure. Whatever the exact cause of death of the many spreading gums that used to grace our countryside, the ultimate cause of rural dieback is the rural economy. Eucalypts are mortal and sheep and cattle are happy to consume their seedlings. In most circumstances, if stock are removed from paddocks containing live eucalypts regeneration will occur and rural dieback will be defeated. The ten percent per annum rate of tree mortality recorded for some of the plains country will continue until no native trees remain in paddocks, unless such regeneration is allowed. Even if fencing of small areas allows regeneration from the seed held on the big trees, the big trees will still die at the same rate. After all, they were trees when the first pastoralists arrived. On the New England Tableland even fencing may be insufficient. Regeneration fails when the remnants are too small to provide habitat for the predators of defoliating invertebrates.

In the semi arid and wet-dry regions of the centre and north and the high mountains of the south-east, grasslands and grassy woodlands, dominated by Kangaroo Grass, survive as extensive communities, perhaps protected from the full impact of stock grazing by their inhospitable climates. In the temperate lowland south-east there are few areas larger than 500 ha where either grassy woodland or grassland survive. Rubbish tips, cemeteries, railroad reserves, road reserves, undeveloped reserves in cities, holding paddocks for stock destined for slaughter, horse paddocks, racecourses and vacant blocks contain grassland species that were once common but are now rare. Which species survive in which location depends on the history of management; road reserves having different species complements from the more frequently burned railroad reserves, which, in turn, have different species complements to lightly grazed paddocks. Almost all remnants have a substantial component of introduced weeds.

Development no more respects the last remnants of a vegetation type than it does a type of bush about to be cleared. For example, the grassy woodland at the site of the new Australian Parliament House in Canberra was one of the few known extant localities for an endangered daisy species (*Rutidosis leptorrynchoides*). Road widening, using herbicides on roadside growth and occupying of 'waste' land for community facilities are all contributing to the steady attrition of the pathetic shards of a once extensive ecosystem.

The virtual disappearance of the temperate lowland native grasslands and grassy woodlands illustrates the often highly selective nature of bush destruction. The same environmental factors that enable the establishment and survival of different types of bush also affect the range of possible agricultural activities, the economic feasibility of agricultural systems, and their productivity.

TECHNOLOGY AND BUSH DESTRUCTION

Australia has many difficult environments for European-style agriculture. The utilisation of these environments has depended upon technological innovations. These innovations have created massive surges of clearance in particular vegetation types.

The Commonwealth Scientific and Industrial Research Organization (CSIRO) was established in the interwar years to provide a scientific basis for development. The Land Research Division of CSIRO have effectively used vegetation as a means of predicting the agricultural development potential of much of northern Australia and New Guinea through their extensive land systems studies. In these studies they identified land units classified according to the physical and biological environment, and assessed their capabilities for development as well as the hazards involved. The same organisation has been directing research towards increasing the range of agricultural activities that can profitably be undertaken within any land system.

The work of CSIRO scientists has provided information that may soon lead to a transmogrification of much of the northern Australian bush through the cultivation of an introduced shrub belonging to the pea family. *Leucaena leucocephala* is easy to grow and provides excellent tropical fodder elsewhere, but has proved poisonous to Australian stock. The CSIRO scientists have identified a bacterium, now absent in Australia, that enables cattle to digest the plant safely.

A seemingly equally innocuous discovery led to the clearing of vast areas of heath on deep sands in the 1950s and 1960s. Heath is a vegetation type that is so well adapted to an extreme lack of nutrients that many of its constituent species die when treated to a soil of normal garden fertility. Small, hard and prickly-leaved shrubs less than two metres tall dominate heath, which usually contains a wealth of higher plant species, sixty having been recorded from one ten by ten metre plot in Tasmania. Many of these species are spectacular in flower, and some are now being grown in Western Australia for the European cut flower market. Heath occurs on some of the poorest of sites in the wetter parts of Australia. Plant nutrients are always in distinct poverty, but the exclusion of trees can be due to any combination of a high fire frequency, extreme exposure to salt spray, soil drought or soil waterlogging.

Before the CSIRO discovery, the continuing economic uselessness of heath seemed so assured that there was little hesitation on the part of State governments in making the most scenic areas containing it into National Parks. Outside the National Parks, some of the more extensive areas of heathland in south-eastern Australia were used for the rough grazing of stock, particularly where small, interspersed areas of other vegetation types enabled the sheep or cattle to gain some sustenance. The Ninety Mile Desert of South Australia was one such area. The poor land and the depression of the 1930s combined to impoverish many of the farmers in this area who, instead of buying food, resorted to living off their own stock. An epidemic of coast disease amongst the farming population was the direct consequence of this change of diet. The victims, ignorant of this fact, were concerned that they were now suffering a disease previously confined to their stock. CSIRO research in the late 1930s and early 1940s identified the cause of coast disease in both stock and humans. It was a cobalt deficiency, the trace element being essential in some small amount for the adequate performance of both stock and human, and being distinctly lacking in the soils, and in the plants on which the stock fed.

The identification of this micronutrient deficiency precipitated research into means of establishing pasture on the deep, sandy soils of the South Australian south-east. This research led to the development of a simple and cheap procedure for converting heath into improved pasture, the Ninety Mile Desert became the Coonalpyn Downs, and the heath became no more. One million acres of heath a year was converted to pasture by the proud developers of the marginal sand plain country inland from Esperance in Western Australia. Most suitable heathland was developed by the end of the 1960s, but conversion still continues in places such as Woolnorth in the north-west of Tasmania, where wet heaths are drained, ploughed and fertilised in a production line approach, at the rate of several hundred hectares each year.

Farming the deep sands is a form of hydroponics, pasture maintenance being dependent on continued inputs of the superphosphate and trace element mixture that allows the nutritious introduced grasses to survive. The future of deep sand farming is therefore only as secure as the price and reserves of rock phosphorus and guano.

The above are but two of many examples of the impact of technological innovation on the survival of the bush. It can be concluded that there is no long-term safety for any type of bush in contemporary economic uselessness.

TREE FARMING

The south-east of South Australia has often been a leader in rural developments. Part of the reason for this undoubtedly lies in the poverty of the South Australian agricultural resource base. The Victorian – South Australian boundary can be easily seen from the air, land that the Victorians regard as not worth developing being the prime land of the state one step away. South Australian leadership has been characterised by excessive optimism, exemplified by the 1880's theory that rainfall follows the plough. This theory encouraged an extension of wheat growing into land that received too little rainfall to ever allow success.

South-eastern South Australia once supported some considerable areas of eucalypt forests and woodlands. Large areas of these native communities were cleared in order to grow an organism that has been fondly referred to as the miracle tree. The miracle tree is Radiata, or Monterey, Pine (*Pinus radiata*), a species highly familiar to all Australians and New Zealanders but so rare as to be almost endangered in its native United States. There is one highly popular and very small peninsula reserve near Monterey in California where, for a small fee, Radiata Pine can be seen growing with the almost as popular Monterey Cypress, in its natural habitat on land not in danger of development. Well-fenced and well-surfaced paths lead the numerous visitors among groves of pine and cypress under which can be found an attractive profusion of other Californian native trees, shrubs and herbs. The trees are far from straight and spirelike, bearing more resemblance to the spreading gums that the pine replaced in South Australia than to the pine in its colonising role.

The Radiata Pine grown in Australia and New Zealand has a carefully selected genetic inheritance that gives fast and straight growth, good self-pruning and excellent wood quality. This selection has been possible because the Radiata Pine is genetically highly variable. This high variability makes the species better able to survive in the rapidly changing Californian environment than if it were more uniform in its constitution. The death of many young individuals of Radiata Pine as a result of being unsuited to a particular site does not matter because it is a species, like most of the eucalypts, that depends largely on fire for regeneration opportunities. Masses of seed are released from cones after a fire kills or damages the adult pine, and there is no space for all the subsequent seedlings to survive to adulthood. If early death is the inevitable fate of most seedlings, high variability holds no disadvantages.

The first planting of Radiata Pine on the sandy soils of south-eastern South Australia produced excellent crops. The pine put on volume at a rate that far outstripped any contemporaneously established eucalypt seedling and did so on a poor site for eucalypts. The second crop was not as good, and the third crop was so poor that it became the focus of considerable scientific research. This productivity loss may have been largely a virgin land effect, as is found with most agricultural crops. However, the mere presence of pines leads to a deterioration in soil quality when they replace vegetation types producing less acid litter. Pines promote a process called podsolisation, whereby plant nutrients, iron and aluminium are carried down and often out of the soil profile after having been made mobile by the various acids released from the foliage and the tree litter. The pines appear to be more effective in promoting podsolisation than the eucalypt communities they usually displace.

Almost one million hectares of native forest has been felled and replaced by plantations. Over 90% of this area has been planted to various species of pine with most of the remainder being dedicated to eucalypts. Only a small proportion of the total planting has taken place on the farmland abandoned after the ill-conceived clearance of wet forest on steep land in the late nineteenth century. Some of the bush that has been destroyed for pines has died in vain. For example, a large area of pines was planted on the coastal sand dunes behind Ocean Beach on the west coast of Tasmania. This plantation has been written off and is included in the 10% of Tasmanian pine plantations that have failed because of the use of unsuitable sites for even the first rotation.

Pine planting continues at a rate of tens of thousands of hectares each year, while planting of fast-growing provenances of eucalypt species, particularly *Eucalyptus nitens* and *E. globulus*, has expanded dramatically during the 1980s and early 1990s. Both the pine and eucalypt plantations usually require the destruction of native vegetation. Of course, many of the native understorey species will survive in eucalypt plantations, whereas only a few ferns survive the mature Radiata Pine forest, and, in a few cases, the eucalypt species planted are the same as those previously cleared from the site.

Clearing native forest to plant Radiata Pine and exotic eucalypts might have destroyed another potential miracle tree. Although this is probably not the case, as scattered remnants of the original tree species can still be found in most areas, the fear of losing such trees has led to a nationwide program aimed at identifying and conserving the genetic variability of *Eucalyptus*. It is likely that the future for wood production in Australia will lie in plantations of genetically improved native species. Care will have to be taken to ensure that their establishment does not eliminate any other elements of our native biological diversity.

URBANISATION

The sprawling Australian city has displaced much less bush than either farm or industrial forest, even when a fair proportion of the area that has been inundated for water supply and electricity is credited to the urban account. As the Australian cities have grown they have incorporated small areas of uncleared bush along coasts, in river valleys and on hills while most of the landscape has been converted to roads, buildings, green lawn, and exotic shrubs, trees and flowers (Plate 4.2). From the 1950s onwards most of the large Australian cities started to reach the bounds of their potential water supplies, and it became less and less desirable for the suburban householder to maintain green lawns and moisture-profligate exotics during the aridity of the Australian summer. The expense of water, and the frequent imposition of inconvenient restrictions on its use, made many people ready to adopt Australian natives in place of the lusher and greener plants of Eurasia. Houses were built amongst the bush, rather than all the bush being erased. Thus, the urban forest has developed a substantial native component, and the bush remnants in suburbia were, for the first time, thought by a substantial proportion of the city population to have a high value in themselves. Some of the rarest plants in Australia are now confined to bush remnants in cities, having been wiped out elsewhere by the practice of agriculture.

Plate 4.2 The bush remnants and gardens of suburbia at Black Rock, Melbourne, Victoria. Most of the bush cover survives in the roughs of golf courses and the narrow coastal reserve. Note the rubbish dump on the cliffs of Red Bluff (N. Rosengren).

THE CHARACTERISTICS AND ECOLOGY OF AUSTRALIAN CULTURAL VEGETATION

Cultural vegetation has been created by humans and largely depends on our inputs of energy and materials for its continuing existence. The most dependent vegetation tends to be the floristically simple grass and herb crops, although the polymorphic cabbage, and the humble potato are capable of the occasional escape.

The Australian agricultural ecosystem is one of deliberate simplification. Herbicides, pesticides, hormonal stimulators and fertilisers are employed to maximise the production of a single crop over a large area during any one time period. The energy from irreplaceable fossil fuels that goes into this process exceeds the energy that emerges from it. Exceptions to this are sugar cane production, where the wastes, bagasse, are used to provide energy, and the wheat – sheep belt, where the rotation of grain crops with nitrogen-fixing legumes uses nature to provide the fertiliser.

The Australian capital-intensive monocultural systems seem unlikely to be sustainable on any but the best soils on gentle slopes, just on the grounds of the accelerated erosion they cause. The methods of commercial agriculture labelled as organic, because they use chemicals formed in natural processes rather than chemicals formed in factories, do not necessarily promise a more sustainable cultural vegetation. Much commercial organic agriculture depends on clearing new bush for fertility and freedom from pests. Permacultural organic systems are theoretically promising, as they mimic the balancing complexity of natural ecosystems. However, their widespread use would require a change from the current socioeconomic paradigm which promotes short-term profit-taking rather than long-term sustainability.

The crop plants that comprise the Australian agricultural landscape are more restricted in number than the native plants used by Aborigines, and they are almost without a native component. The Macadamia Nut tree is the only native plant that is cultivated on a large scale in Australia, and its cultivation was initiated elsewhere. These crop plants were accompanied by coevolved ruderals which also depend for their existence on the maintenance of human cropping practices.

Whereas cropland is floristically simple, urban areas have some of the most floristically rich vegetation in Australia. Over three hundred higher plant species can be found on some suburban blocks. These include highly bred cultivars, such as those of the rose (*Rosa*) or *Rhododendron*, non-invasive ornamental species from other countries or regions of Australia, such as the Norfolk Island Pine (*Araucaria heterophylla*) or the Fern-leaf Mahonia (*Mahonia lomariifolia*), invasive, often self-established, exotic ornamental species, such as Blue Thunbergia (*Thunbergia grandiflora*) and South African Boneseed (*Chrysanthemoides monilifera*), weeds, such as Catsear (*Hypochaeris radicata*) and Sensitive Plant (*Mimosa pudica*), local native ornamental plants, and local native survivors of development.

Social status seems to be strongly associated with the structure and floristics of suburban vegetation. Tree cover is strongly positively associated with income within any one urban complex, while the brightness of flower colour shows the reverse relationship. The age of the suburb and its inhabitants is also critical. Garden fashions change over time and vary between subcultures and socioeconomic groups. Young professional couples in the 1970s tended to favour the native garden, which, in the eastern states, largely consisted of Western Australian plants, and in Perth largely consisted of eastern Australian plants. Today, the same group of people tend to grow the classical English cottage garden. The environment provides some limits to the floristic composition of gardens. These are usually imposed by pest and disease problems, especially those that make ornamentals unsightly.

Vacants lots, roadsides, rail reserves, old walls, rubbish tips, parks and gutters all have distinct types of vegetation and flora that are not seen outside urban complexes. Their distinctiveness relates to the nature of the disturbance regimes experienced in different city habitats, combined with the city environment, which is characterised by air full of grime, acid mist and rain and a high input of nutrients.

Cities are nutrient and material sinks. Only air pollution and sewage tend to escape their bounds. Half the agricultural chemicals purchased in Australia are used in the cities. Products of our mines, such as lead, cadmium and zinc are processed, used and released in cities, rendering the produce of some urban soils and estuaries potentially toxic to people.

In the future, the cultural vegetation of Australia is likely to become more similar to the natural and semi-natural vegetation of Australia than it is today. The variable climate and poor soils of Australia make it difficult in the long term to grow many crops developed on other continents. We are likely to see more crops selected from native plants that are preadapted to these conditions. Australian cities are reaching, or have reached, the limits of their water supplies, half of which are now expended on gardens. Native grasses, shrubs and trees are likely to largely replace the water-demanding European plants in gardens and amenity plantings. Although not all exotic plants require large inputs of moisture, the native plants attract native birds to the city and affirm a sense of Australian and local identity. There are already large businesses devoted to collecting seed and propagating plants for ecological horticulture — growing and establishing local native plants. Native grasses now grow in St Kilda median strips. At the same time as the city vegetation is becoming more pastel in its tones, the bush is incorporating more blue-green exotics that escape from the ornamental pool.

THE CONVERGENCE OF CLEARANCE AND RESERVATION

The area of National Parks and other reserves of similar status has increased in Australia at a rate that would more than please any board of directors if it were their profits. In 1971, 0.56% of Australia's area was protected from development by Acts of Parliament. In 1989, 5.4% was reserved, purportedly in perpetuity, for the benefit of native animals and plants and the people who wish to ensure their survival and enjoy their existence. Yet, the expansion of the National Parks systems of Australia has seen no abatement of the rate at which bulldozers, fire and floods are destroying natural vegetation (Figure 4.2).

A large proportion of the surviving wet country bush is on private land. Only the South Australians and Victorians have attempted to control private land clearance for conservation purposes. In most states Crown Land is still being alienated for agricultural development, and the rate of bush clearance on private land is still extremely high and extremely selective. Analysis of satellite images has shown that 9000 hectares each year were cleared in Tasmania in the 1970s and 6000 hectares each year in the 1980s. The rate at which some types of poorly-reserved and threatened forest are being cleared rivals the 1% per annum recorded for the Amazon rainforests. Unless clearing is directly controlled on private land it is inevitable that almost all private land will be cleared, even on slopes and soils totally unsuited for any sustained production. If only one landowner in one hundred is motivated to clear all the bush on his or her private land, it will take not much more than one hundred generations before almost all such bush is cleared.

It is apparent from the most casual observation that many more than one in a hundred landowners are very much engaged in a campaign of total eradication of their own bush even where that eradication is economically as well as ecologically irrational. The destruction of bush on poor sites has been accelerated by the rural retreat phenomenon. Rural retreaters are urban dwellers with rural fantasies that they attempt to materialise by establishing peasant farms in the bush while living off money gained in the city. Blocks of poor land can be profitably offloaded on to these people, who will develop their properties without the necessity of profit. The irony of rural retreating lies in the destruction of the bush by some of its most romantic devotees.

The large areas in the National Parks systems fail to protect the full range of native ecosystems and species. For example 13% of the vegetation types mapped for Australia at 1:5 million do not occur in any reserves, nor do almost half of the rare and threatened higher plant species of Australia. This lack of representativeness has resulted from the bias towards scenic grandeur and economic uselessness in the selection of land for National Parks and from the great expense that would be involved in resuming the few areas of bush that survive on our best agricultural land. Information on the nature of the gaps in State Reserve systems is still incomplete at any but the crudest level for most vegetation types. Inventories of species and communities are unavailable for many of our most famous parks, much less for the park systems of whole states. However, there is no doubt that temperate and tropical tussock grasslands, temperate grassy woodlands, Brigalow, saltbush shrublands, mulga scrub and the tall wet forests dominated by Mountain Ash and its close relatives are seriously under-represented in reserves. Vegetation types like heath and rainforest are only well-reserved when viewed as a whole — detailed studies of the distribution of their major variants have invariably shown that at least some are unreserved.

Alpine vegetation is the best reserved of any of the major Australian vegetation types, yet one extreme of the plant community and species complexes found in Tasmania is unreserved, and harmfully grazed by stock, as is much of the alpine vegetation of Victoria.

Despite their shortcomings, the National Parks systems of the states and the Commonwealth of Australia have enormous public support and some degree of international and national legislative protection. Many types of bush are well-protected and most of the National Parks authorities are attempting to rectify the gaps in their reservation coverage as money and political opportunity allow. It becomes more difficult to reserve all types of bush as more variants of the different vegetation types are recognised. However, it has long been asserted in soil conservation that there is more to be gained through investing in the productivity of already cleared agricultural land, than in extending clearance to the poorer lands still under bush. Thus, there is a strong case for stopping any further major clearance of bush, whether publicly or privately owned. The results of the way we have treated our best agricultural land give little confidence in the long-term fate of more marginal lands if developed. Accelerated erosion and salinisation have followed much past clearance and their elimination requires very large investment. In short, it would seem prudent and economically sensible to leave the remaining bush alone, at least until we are capable of husbanding the land we have already transformed.

■

5

The impact of forest use

Although native forest covers only 5% of contemporary Australia (Figure 5.1), it has formed the focus for some of the more protracted and intense development/conservation debates of the late twentieth century. Inquiry after Royal Commission after Inquiry have failed to resolve the basic conflict in values between those who wish to maintain our native forests for non-tangible use and those who wish to maximise timber output.

The massive clearance of forest documented in the previous chapter provides only a part of the picture of the impact of people on native forests. While even the worst of logging regimes has ecological impacts that seem benign compared with clearance and conversion to crop or pasture, only 17 million of the 43 million hectares of surviving native forest have not been logged in the last two hundred years, and much of the unlogged forest has been subjected to changed fire regimes. In addition, approximately 10 million hectares that have previously been unlogged are, at present, available for future logging, and much forest that has been logged selectively in the past is likely to be clearfelled in the future.

This chapter attempts to elucidate the past and likely future impacts on the native flora and vegetation of using forests for wood production.

Figure 5.1 The distribution of forest in Australia (based on the maps in AUSLIG (1990). Atlas of Australian Resources: Volume 6 – Vegetation. AGPS, Canberra, and observations of the author).

A VERY BRIEF HISTORY OF WOOD EXTRACTION

The forests of Australia have been cut since the earliest decades of settlement in New South Wales, when Red Cedar (*Toona australis*) was exported from the coastal rainforest. During the late eighteenth and early nineteenth centuries there were perceived to be few people and vast areas of forest. These were plundered for their timber, then abandoned to their fate, which was, as often as not, conversion to agricultural land.

As the mirage of endless forests gave way to a perception of the reality of interminable shrublands, spinifex and pasture, State Forests were declared and foresters first imported then trained locally to manage them. By 1990 State Forest covered almost 11.5 million hectares, approximately one quarter of the remaining native forest estate. Another quarter of the remaining forest was privately owned and the larger proportion of the remainder was Crown Land used for commerce rather than conservation.

THE FORESTRY PARADIGM

The basic concept in production forestry is sustainable use, which entails an even or increasing flow of wood from the forest in perpetuity. This flow is theoretically ensured by adjusting cutting to forest growth. Maintaining the flow is made difficult by an inadequate understanding of both current forest productivity and the medium and long-term impacts of use on the productive capability of forest soils. Frequent changes in the nature and magnitude of the market for forest produce make it difficult to plan appropriate rotation lengths and silvicultural treatments. A eucalypt forest managed for sawlogs must be cut on an 80–90 year rotation to allow the trees to get big enough to be suitable for this purpose, whereas a forest managed solely for pulpwood would be managed on a 30–40 year rotation, which maximises wood yield. Thus, if market preferences shifted to sawlog from pulpwood, there would be little medium-term opportunity for response from pulpwood forests. Changes in social preferences for land use necessitate frequent adjustment of calculated total wood flows. Disasters such as the 1983 fires in Victoria and South Australia can dramatically modify sustained yield calculations. On top of all this, the forests that are being theoretically managed for sustained yield are still largely the free good provided by nature and Aboriginal land management. These old trees do not produce as much wood per annum per unit area as rapidly growing young forests, yet cannot be cut all at once if a smooth flow of wood is desired.

The other major conceptual element in the forestry paradigm is multiple use, the idea that areas of forest can be used sequentially or contemporaneously for a variety of purposes. While there is no doubt that there is a wide spectrum of forest uses that can be accommodated in this framework, there are many uses that are incompatible with most or all forms of production forestry. These include wilderness use and some aspects of nature conservation.

THE IMPACT OF SILVICULTURAL SYSTEMS ON EUCALYPT FOREST

Silvicultural systems are the set of techniques and procedures that are used to cut and regenerate forest. Eucalypts were, and still are, the major timber resource of Australia. They include some of the fastest growing trees in the world and species that will grow on the poorest of soils. It is difficult to prevent them from regenerating. If a small group of trees is removed it is likely to be rapidly replaced by previously suppressed saplings and seedlings or by new trees establishing in the disturbed ground.

Australian foresters developed the silvicultural technique of group selection, which partly mimicked the natural gap phase replacement process. This system worked well in most eucalypt forests. For it to work perfectly would have required that the trees of poorest form and character in a group were felled along with those of best form and character. Yet the high degree of outcrossing and genetic variability in eucalypts, and the selection for the genotype most suited to the site that occurs as the trees thin out, reduce the possibilities of genetic degradation. Group selection meant that the multiple age classes present in most eucalypt forests were maintained by the logging program. The forest continued to be recognisable as forest.

Tasmania has the highest proportion of forest cover of any Australian state, almost 40% of the island being covered by trees. A large part of this forest consists of a dense understorey of small broad-leaved or rainforest trees beneath the emergent eucalypts. The group selection system had highly variable success within these wet forests and similar forests on the mainland. A knowledge of the ecology of these forests held the key to regeneration success. Research projects undertaken in Tasmania and Victoria during the 1950s documented the inability of the eucalypts to survive and grow as seedlings in small gaps where the understorey of the eucalypts had a closed canopy. The understorey of much of the wet eucalypt forests is an incipient rainforest. The interlocking canopies of the rainforest trees and broad-leaved shrubs allow less than 1% of the incident solar radiation to reach the ground, and selective logging seldom creates a large enough gap in this understorey to allow enough light for successful eucalypt establishment. The eucalypts germinate but are soon destroyed by fungi.

Thus, the researchers found that the eucalypts are in precarious possession of well-watered land. Most trees had established after major conflagrations had killed both the fire-sensitive adult eucalypts (the eucalypts growing in these areas being among the few incapable of vegetative recovery from fire) and the dense understorey. The seed that dropped from the fire-killed parents after fire was protected from the flames by woody capsules. The seed bed was free of competition, sterilised by heat and fertilised by ash. Eucalypts retained possession of any burned site unless another fire eventuated before the new generation produced seed or unless they died of old age before the next fire. Both these alternatives do occur, the first resulting in vegetation dominated by grass, tree fern, bracken or composite shrubs, the second resulting in rainforest. From the point of view of the eucalypts the safe period between fires is 10–300 years, and the large area of wet forest that greeted the first European settlers must have resulted from long-term frequencies within this range. As the penalties for dropping out either end are a very slow haul back from the edge of the burned area, the frequencies probably largely hovered around the middle of the range. These findings induced the foresters to attempt to mimic nature by clearfelling and burning. All the merchantable trees were removed from an area between 100 and 400 hectares (the coupe), the understorey felled by bulldozer, and then the foresters waited for the perfect day.

The perfect day is one on which the felled understorey and the slash from the felled eucalypts are dry enough to burn, but the surrounding forest is not. There are not very many of them and mistakes are made. If it is too wet to burn no harm is done, but if the fire escapes disaster results, as fires in wet forest are uncontrollable and tend to spot many kilometres ahead of the main fire front. The perfect fire on the perfect day is one that burns in from the margins of the coupe, and forms the convection column and mushroom cloud so familiar to the inhabitants of the nuclear age. Aeroplanes are used to broadcast seed once the coupe has cooled and flying conditions permit. An occasional baiting or shooting of the marsupials attracted to the regenerating forest completes the management procedure.

There is no doubt that this slash and burn silviculture has solved the problem of eucalypt regeneration in the wet forests, which has emboldened most foresters working in commercial eucalypt forest to adopt the same system. Not that there are not some variations on the basic theme. In New South Wales the clearfell areas are much smaller than in Tasmania. In Western Australia, Victoria and elsewhere seed trees have sometimes been left to be removed later, thus saving the expense of aerial sowing.

Most forests to which clearfelling and burning are now applied are not naturally even-aged and certainly do not require such drastic treatment to induce eucalypt regeneration. The clearfell and burn system is attractive to foresters because it is easy to administer, and has some initial economies when compared to the alternative of group selection. However, it is new and relatively untested, considering that the second crop is decades away in the shortest of conceived rotations, and that the oldest clearfell coupes in dry forest were regenerated less than two decades ago. Those in wet forest are not much older.

The impact of the new silviculture on the large proportion of forests to which it is planned to be applied promises to be substantial. Patterns of dominance, structure and floristic composition will change in both wet and dry forest types.

Rotation times in the wet forests are planned to be between forty and ninety years. If these rotation times are used the wet forests will be burned more frequently than was likely to have been the normal case under Aboriginal management of Australia. The burns are also far more severe than most that would have been experienced under natural conditions. Their increased severity is caused by the high fuel loads given by the felled understorey and the heads of the trees whose trunks have been removed for timber. The combined effects of increased frequency and increased severity will be a shift in the understorey vegetation away from rainforest species and towards shrub and tree species that are better adapted to survive fire. Forests of giant eucalypts with understories of rainforest trees (mixed forests), such as the Shining Gum (*Eucalyptus nitens*) forest with a Southern Sassafras (*Atherosperma moschatum*) rainforest understorey found on the Errinundra Plateau in East Gippsland, will become exceedingly rare. In another seventy years these mixed forests will only survive in National Parks and water catchments. The National Park managers will be faced with a terrible problem. The only way that they will be able to perpetuate the giant mixed forests will be to burn them before the eucalypts die of old age. Yet the areas of these forests that they control are so small, and usually so old, that there is little prospect of maintaining all the stages of development of mixed forest at the same time. If they do not burn these forests, or if they are not burned by accident, rainforest will take over from mixed forest. The type of rainforest that takes over from mixed forest is well-reserved and in little danger of extinction. Yet, for the first century after a burn of a mixed forest in a National Park the forest will hardly differ in botanical characteristics from the hundreds of thousands of hectares of forest regenerated for timber.

The continuous perpetuation of natural mixed forests dominated by giant eucalypts can only occur if large areas of these forests, in different stages of their lifecycle, are reserved in different areas. Reserving a hectare or two around particular forest giants has no long-term conservation value.

Some of the logged wet forests may experience a long-term deterioration in soil fertility and soil depth as the result of clearfelling and regeneration burning. The data are incomplete. For almost all Australian forest, there is no exact knowledge, through measurement, of the inputs of nutrients from dust and seawater precipitated in rain, or of the rate of release of nutrients through the weathering of the bedrock, or of the amount of nutrients that are lost in runoff and groundwater after fire. There is information on nutrient loss to the atmosphere in the smoke of the regeneration burns. This proved to be substantial: in the order of one tenth to one fifth of the nutrients found in all the material that remained on the ground surface after logging. There is also some information on the amount of nutrients contained in the logs that are removed elsewhere for processing. This loss is also substantial.

Unless the soils had been steadily increasing in fertility previously there seem good logical grounds for expecting that increased fire frequency with increased fire intensity would lead to a long-term decrease in the soil nutrient store. Studies showing that soil nutrient stocks are greater after regeneration burns than before cannot be interpreted to mean that long-term site fertility is increased by firing. The increased nutrient status of the soil after fire occurs contemporaneously with a marked decrease in the total amount of nutrients held in the soil and vegetation combined. The vegetation has been robbed to feed the soil, the basic principle of slash and burn agriculture as well as of slash and burn silviculture.

A drastic reduction in the nutrients available to the plants in these wet country ecosystems would result in a change to the species composition of the vegetation. This change would favour slower growing and more small and hard-leaved species than those occupying the sites at present. However, these types of changes may not occur because the response of the timber trees to any decline in nutrient status is likely to be be a decline in growth rate. Thus, trees will have to be allowed to grow longer in order to get the same size of log. Such a forced increase in rotation time would allow inputs of nutrients to compensate for the losses at establishment.

Accelerated soil erosion may be a greater cause of vegetation change on some of these mixed forest sites than any decline in nutrient status. Forests are felled on extremely steep slopes and on shallow limestone soils. In these cases the loss of soil following clearfelling and firing is likely to be much greater than ever occurred with natural conflagrations, which lacked the disturbances caused by machinery. The situations in which clearfelling and regeneration burning will lead to a reduction of the physical soil resource are yet to be established, although the unfortunate results at the extreme sites discussed above are easily observed.

In most wet forests the seeds sown on the coupe after burning are from the same eucalypt species that grew there previously, and often are from the best of the felled trees. However, in dry forests, eucalypt species that have little or no commercial value will be omitted from the seed mixture, and economically more desirable species, not naturally present on a coupe, will sometimes be included. Whether these attempts to change the forests will succeed in producing almost pure stands of commercial species is uncertain. The non-commercial species often occupy the driest or most ill-drained parts of coupes where commercial species cannot survive in the long term. Also, the amount of regeneration that actually results from the seed sown by foresters varies enormously. Many suppressed seedlings survive regeneration burns, and they often constitute the bulk of the new eucalypt growth. Nevertheless, some eucalypt species are having their ranges extended beyond their natural bounds, both within the small area of a coupe and through the State Forests as a whole. These local and regional range extensions may result in an increased incidence of hybridisation, as species that are not normally juxtaposed are sown in mixture. The prospect of eucalypt forests consisting of complex hybrid swarms between two or more species is one that has already been realized in second-generation eucalypt plantations in California. Thus, the fine adjustment of species to site that is so apparent in the natural forests may be destroyed.

In some cases there is a danger that timber production values, as well as conservation values, could be reduced by the homogenisation and commercial bias of seedfall within clearfell coupes. Research undertaken in the Mt Lofty Ranges near Adelaide, South Australia gives some indication of possible outcomes. In this area stands of Messmate Stringybark (*Eucalyptus obliqua*) are confined to the south-facing slopes of gullies, while peppermint and gum species dominate the drier sites. Given a reasonably constant supply of moisture the Messmate will grow faster than any of the species found on adjacent drier sites. The dry site species are cautious in their use of water. They close their stomata at the first hint of soil drought, thereby ensuring that they conserve moisture, but at the cost of cessation of growth, as no water released through stomata means no carbon dioxide gained for photosynthesis. Messmate keeps its stomata open virtually no matter what happens to soil moisture, thereby ensuring either continued rapid growth or sudden death by thirst. If the remnants of bush in the Mt Lofty Ranges were to be logged, clearfelled, burned and sowed with the seed of the most economic eucalypt species, Messmate would dominate the seed mix. It is one of the ash group of eucalypts, the species in which are highly prized for both cut timber and pulpwood. With most competition for moisture destroyed by the regeneration burn, the Messmate seedlings would probably establish on all slopes and aspects, and outgrow and suppress much of the regeneration of the other species.

However, once the regenerating forest developed enough leaf area to use all of the limited summer moisture resources, considerable dieback of Messmate could be expected. On some sites Messmate might grow happily for decades, then suffer mortality in a rare sequence of dry years. Death of adult Messmates at their dry natural limit has been recorded for Victoria during the drought that peaked in 1968 and for Tasmania in the 1983 drought. In both cases the species from drier sites survived quite easily in mixture with the dying Messmate.

The significance of the species mixture problem lies in the dramatic differences in available moisture that occur between closely adjacent north and south-facing slopes in middle latitudes. In an area at 43°S with a mean annual rainfall of 500 mm, the slopes most sheltered from the evaporative effects of the sun have the wet forest vegetation of a flat area with 900 mm of rainfall while the slopes most exposed to the sun have only sufficient available moisture to support a mallee woodland.

Some of the effects of clearfelling and burning on dry forest will only became apparent after each coupe has been cut several times. At least some old trees with the hollows so essential for the breeding of many birds and larger animals have been left in many clearfell coupes. However, their lives are limited and there is no prospect of their replacement, as the trees now growing are destined for industry at a relatively tender age. The birds and animals that require these old trees, and those that require the varied resources of a forest consisting of many size classes of trees, will decline in abundance, while species adapted to the new conditions will be favoured. These shifts in the animal population will certainly lead to vegetation changes that are virtually impossible to predict, as we know so little about the functioning of the dry forest ecosystem and the role of various bird and animal species in plant dispersal and pollination.

Some environmental changes that seem highly unlikely to be caused by the new logging and regeneration practices are those that worry the most people. There is a genuine fear that felling forests lowers rainfall (the reverse concept to rainfall following the plough). Where trees act to trap and coalesce small droplets of mist that would otherwise move elsewhere unprecipitated, there is no doubt that their elimination decreases rainfall. However, in most forests most precipitation is received because air rises, in rising it cools, and in cooling it cannot hold all its moisture as vapour. Air is forced to rise over the mountains on which many forests grow, it rises in the disturbances generated by passing fronts and it will also rise if one area of land is hotter than another because it reflects less solar radiation. The last mechanism is aided by patchwork clearfelling, while the first two are unaffected.

However, clearfelling can have extreme effects on the quality of water in streams. The machinery used in clearfelling exposes the soil to erosion, which can increase stream turbidity, especially where there are no protective buffer strips of uncut bush. However, the major impact of clearfelling is in those regions where salt is found in the subsoil or groundwater. Trees transpire enormous amounts of water. If they are eliminated from large areas, more water is likely to remain in the soil, and therefore more water will move through the soil into streams. The increased groundwater movement from some overcleared parts of the Great Dividing Range in Victoria has been the ultimate cause of the salinisation problems experienced in some irrigation areas. Salinisation affects riverine vegetation as well as the farmers and townspeople downstream.

Continued logging of Australia's eucalypt forests will undoubtedly change aspects of their structure, floristics and productivity. However, the resilience to disturbance of most species in eucalypt forests makes them far less susceptible to basic changes as a result of logging than is the rainforest ecosystem.

The impact of rainforest logging

Eucalypts and Radiata Pine account for most of the forestry activity in Australia today. However, some of the woods of highest value per unit volume were extracted from rainforest. The rainforests of Australia have had about one third to one half of their area destroyed for farmland, and much of the remainder has been favoured for inclusion in National Parks. However, much rainforest is still found within State Forest and uncommitted Crown Land, especially in Tasmania. These forests have been logged selectively for speciality woods.

The tropical and subtropical rainforests have suffered most of the land clearance. These rainforests are characterised by the large number of species of tree that can be found within small areas, up to 164 tree species in one tenth of a hectare being recorded for one of the richest variants. In its natural condition the forest has reasonably constant populations of its constituent plant species, the death of an individual of a species being compensated for by the establishment of another individual of the same species elsewhere in the forest. Individual large trees can die naturally of many causes, some, like wind-throw or landslip creating large gaps in the canopy, and some, like fungal attack, killing only the occasional tree. There is a complement of light-demanding, fast-growing and short-lived tree species that quickly occupy such gaps, the species composition of the parts of the forest consisting of these secondary species varying according to disseminule availability, gap size and other environmental conditions. Long-lived, slow-growing, shade-tolerant species can also establish early and eventually replace the secondary species. These primary species include many of the best quality timber trees. Which primary species establishes where depends on chance as well as environment, and chances are likely to be improved as the numbers of adult individuals of any species increases in an area.

These attributes of the dynamics of the tropical and subtropical rainforest make maintaining the numbers of particularly favoured species fairly difficult. Selective logging drastically increases that part of the forest dominated by secondary species and thereby increases the chances of these species establishing in competition with primary species. The difficulties involved with sustained yield of desirable species in tropical and subtropical rainforest have not been overcome in Australia; the long gaps between cuts, and the restraint that would have been necessary for any hope of sustained yield of desirable primary species, ensured that these species were mined and abandoned, or replaced by the more gregarious Hoop Pine (*Araucaria cunninghamii*) wherever the economics of plantation establishment were perceived to be reasonable.

Although most of the species that constitute tropical and subtropical rainforest could not survive the fire regimes that are experienced in the adjacent encalypt-dominated forests, these types of rainforest seldom burn. They are protected at their margins by species whose foliage is non-inflammable and fire-suppressive. The heat and high humidity that is the general condition on the forest floor encourages rapid breakdown of litter and the forest canopy allows through too little light for a ground stratum of any great density to develop. Thus, there is little or no accessible fuel for a fire. It is only when forests are broken up by felling or cyclone that sufficient dry fuel is found in the ground stratum to support a fire. In most natural situations fire can only lead to a slow marginal retreat of the rainforest.

The temperate rainforest is concentrated in Tasmania where it extends on to some of the coldest, wettest and most infertile ground in Australia. Consequently the breakdown of organic matter does not keep pace with its accumulation, and rainforest is widely found growing on red fibrous peats. These peats will burn during the more extreme dry spells of summer, and once combusting are only extinguished by prolonged and soaking rain. Little of the temperate rainforest has been cleared for agriculture, but it covers only one tenth of the area it could cover if fire were absent from the Tasmanian environment. The peat fires cause the death and collapse of trees, creating a jumble of wood ripe for fueling the next fire.

The destruction of the temperate rainforest by fire seems to have accelerated in the last thirty years. The vegetation map of Tasmania shows 56,000 hectares of recently burned rainforest compared to 166,000 hectares that was burned 30–200 years ago and is now dense scrub, and 682,000 hectares that is still rainforest. As men light virtually all fires in Tasmania there is no hope for a decrease in this rate of attrition unless the incidence of incendiarism and escapes from authorised fires drastically decreases. The inhabitants of the mining towns of western Tasmania are so prone to 'go for the match' that government authorities feel that they are forced to conduct a program of appeasement burning of the treeless plains, the locals openly threatening to light fires unless the authorities do it for them.

The west coast loggers of King Billy Pine (*Athrotaxis selaginoides*) used to burn patches of rainforest in order to have somewhere safe to put their bulldozers. Some of the rainforest they were logging was dominated by trees more than 1000 years old. As King Billy Pine is easily killed by fire and is a poor disperser in from fire edges, they had a fire break around them that had worked for at least a millennium, but their logic may have related to the fires lit in hot dry weather by other bulldozer drivers, who were scarring the adjacent hills to expedite the search for mineral riches.

King Billy and Huon Pine (*Lagarostrobus franklinii*) have been assiduously mined from the Tasmanian rainforest since the early years of the nineteenth century. The lode is now almost exhausted with few stands still available for logging and with a 500-year wait for the next crop, if seedlings replace cut adults and if no fire destroys the forests in the interim.

There is little demand for the wood of the Myrtle Beech (*Nothofagus cunninghamii*), Southern Sassafras (*Atherosperma moschatum*) and Leatherwood (*Eucryphia lucida*) that constitute most of the Tasmanian rainforest resource. So most temperate rainforest is now (1993) free from any logging. Yet, the major proportion of its area is within State Forest, or Crown Land that could easily become State Forest. If a pulpwood market ever developed for rainforest tree species there is a very strong likelihood that the rainforest on reasonable soils would be clearfelled then sown to eucalypts, as eucalypts produce much more wood per unit time per unit area than any of the rainforest species. Even if rainforest was retained, silviculturalists in the Tasmanian Forestry Commission have suggested that clearfelling might be the best approach for the Myrtle Beech forests.

Large numbers of the older Myrtle Beech trees have been dying during the last two decades. Even in areas remote from roads or human settlement, some myrtle forests are dotted with dead and dying crowns. The incidence of this dieback increases wherever the root systems of trees have been disturbed and on newly exposed forest edges. Thus, even selective logging destroys more than the trees removed for timber. The cause of Myrtle Beech dieback is a fungus. Only very young trees appear totally immune. The dieback is thought to be a natural phenomenon magnified by environmental change. Thus, its future, and the future of the Myrtle Beech forest, is somewhat difficult to predict.

There does not seem to be any simple way of maintaining the biological diversity of rainforest while sustaining a significant output of wood products. It is fortunate, therefore, that this activity is almost extinct in Australia. The past impact of logging and clearing on Australia's rainforest has been substantial. Unlike the forests of eastern and southern Australia, the forests of the monsoonal north have suffered little from these causes. The next section considers their responses to this minor degree of disturbance.

IMPACT ON FORESTS OF THE WET/DRY TROPICS

In the monsoonal tropics there are also numerous small areas of rainforest situated anywhere where fire has difficulty gaining access. Although these monsoon rainforests cover a very small total area, they are well dispersed throughout northern Australia and the rainforest tree species have even managed to colonise the rubble of abandoned settlements. Many of the monsoon rainforest tree species have fleshy fruits that can be dispersed long distances by the prolific tropical birdlife. However, the rainforests can supply little or no timber, as most of the high rainfall country of the north is covered by forests and woodlands of eucalypts. These eucalypts also promise little joy for the wood industries, although if they were straighter and more rot-resistant they could be used as pipes, for the larger stems are usually hollowed by termites. The northern Cypress Pine has engaged most silvicultural interest, but many stands have been destroyed because of changes in the frequency and intensity of fire. The survival of the Cypress Pines can be made precarious by a severe fire followed by another fire before the seedlings from the fire-killed adults have themselves produced cones. Fire protection and the inevitable occasional severe fire it occasions replaced the previously frequent but low-intensity burning of the Aborigines. This well-intentioned change in firing patterns has led to dense thickets of Cypress Pine where the severe fire has not been followed closely by another, and the local extinction of the species where the second fire occurred.

SOME ECOLOGICAL ARGUMENTS RELATED TO FOREST LOGGING

Most of the arguments about logging are related solely to the values of the protagonists in the conservation/development debate. However, there are several ecologically based arguments that deserve some consideration in this book.

The future of 'old growth' forests has become a focus of much definitional argument. The conservation movement has promoted the protection of old growth forests from logging for many reasons including the preservation of wilderness, the maintenance of ecosystems largely unmodified by people and the protection of species dependent on large trees. The definitions emanating from the movement tend to concentrate on the relatively pristine nature of old growth forests, rather than on the size class distribution of trees, which has been emphasised by the logging lobby. In a judgement worthy of Solomon, the Resource Assessment Commission suggested that old growth forests should be dominated by old trees and be largely undisturbed by human activity. A suggested policy is to protect such forests. This policy will protect neither old trees nor undisturbed forest in the longer term, as the maintenance of a stock of old trees requires the establishment of young trees, but once the trees in an area are young they will be available for logging. Thus, the conflagration that, in natural conditions, would kill many or all of the old trees and enable the regeneration of their replacements, will convert a protected, undisturbed forest into an unprotected, undisturbed forest.

Some forestry ecologists have suggested that the clearfell-slash burn-aerial sowing regime is ecologically innocuous as it mimics the natural regeneration cycle. The removal of stems, the hotter fires related to the concentration of fuel on the ground, changed frequency of fire, the ground disturbance caused by machinery and the alien species introduced on that machinery all make slash and burn silviculture an ecologically distinct event.

Some proponents of logging have also argued that the disturbances created by logging are ecologically necessary to replace the predation and disturbance of the extinct megafauna, and are invaluable in that they create the intermediate disturbance that is associated with the highest levels of species richness. Given that the megafauna became extinct well before most of the natural forests of Australia expanded from the refuges (refugia) they occupied during colder and drier times, the former argument is poor. The disturbance argument is superficially attractive. There is definitely an increase in species richness in the early stages of regeneration after logging. However, natural disturbances have provided for the survival of early successional species in the past and could do so in the future. Also, an increase in the area occupied by early successional stages could be expected to disadvantage later successional species.

The conservation lobby has argued that forest logging threatens native species with extinction, to which the logging lobby replies that no native higher plant species has become extinct because of logging, and that there is little or no evidence of major floristic changes occurring after logging and burning of eucalypt forest.

The long-term effects of current silvicultural regimes on floristic composition are uncertain, but the spread of Cinnamon Fungus (*Phytophthora cinnamomi*) on logging machinery is a major threat to the survival of native plant species. The localised loss of rainforest species in Tasmania does not threaten their future, given the high proportion of this vegetation type, and the populations of its constituent species, in secure reserves. Given that our forests cover a larger area now than during the colder conditions that prevailed through most of the Quaternary, it would seem theoretically possible to maintain all forest species while logging part of the forest. This would require protecting refugia and pathways to refugia, as well as protecting all communities, species and genotypes in secure viable reserves.

Conclusion

The future of Australia's native forests may be secured in the next few years by an expansion of the forest reserve network and strong controls on clearance and logging. The recently ratified National Forest Policy, although ambiguous in many of its clauses, leads in this direction. However, the potential for further degradation of Australia's native forest vegetation through unsustainable logging activities cannot be denied. A precautionary approach is needed to ensure the future of this major Australian vegetational asset.

6

Fire

Fire is as natural as rain, although fires caused by thunderbolts from the sky are, fortunately, less frequent than their elemental opposite. Active volcanoes are somewhat lacking on the contemporary Australian landmass and the spontaneous combustion of naturally rotting vegetable matter is an extreme Antipodean rarity, so today lightning and people are the main causes of fire. Most discussions of the origins of patterns, processes and biological pathways in the Australian bush require the mention of fire, which influences and in turn is influenced by both vegetation and soils, and which also varies in its incidence according to climate and the propensity of the human inhabitants of an area to light fires. Changes in the type and incidence of fire have led and are leading to massive shifts in the nature and composition of the Australian bush.

A BRIEF HISTORY OF POST-INVASION FIRE REGIMES

Charcoal records in sediments, charcoal incorporated in tree rings and historical records suggest that fire frequency and intensity in forests generally increased after European colonisation. After many decades of accelerated incineration, fire suppression became the usual form of management. This resulted in less frequent, but more severe fires. In recent decades, frequent-low intensity, planned fires have been used in much of the remaining forest estate.

Over most of the rest of the continent, the size and intensity of individual fires, but not their frequency, seem to have increased as a result of the destruction of the Aboriginal socioeconomy and the activities of the European invaders. However, in some areas where stock grazing has been extremely intense, fire has become a rarity through lack of fuel.

FIRE-SUSCEPTIBLE VEGETATION

Despite the reputation of Australia as a continent of smoke, there are types of vegetation on the continent and its associated islands that are easily eliminated by one or two fires because of the incapacity of the dominant species to survive fire or to regenerate or reinvade after fire. Much rainforest fits in this category as do some of the alpine and subalpine plant communities of south-eastern Australia and Tasmania. Much of this vegetation has survived the successive rigours of the aridification of Australia in the Tertiary, the Aboriginal invasion in the Quaternary and the human population explosion in the last two centuries, because it occurs in situations where fire could never, or very rarely, be ignited or sustained, or because, like some tropical rainforest, the vegetation will not burn.

Between 1960 and 1980, seventeen percent of the area of alpine and treeless subalpine vegetation in Tasmania was burned. Two of the smallest fires may have been caused by lightning — most were escapes from burns by farmers, forest regeneration burns, campers' fires, or were lit for the joy of seeing bush burn. Fires burned more of the eastern and western high country than the central high country, which is protected from the spread of fire by deep valleys, high cliffs, glacial lakes and subalpine rainforest, and much of which is remote from the roads that carry most arsonists. Where fire is prevented from spreading by natural features such as water or rock, or prevented from igniting by the fire suppressive geometry and chemistry of the vegetation or its constant wetness, survival of fire-susceptible vegetation types has been better than where survival depended on the absence of ignition by human beings.

Deciduous and coniferous shrubs dominated a high proportion of the alpine and subalpine vegetation that was burned. These species have few or no adaptations that could ensure their perpetuation after an intense fire. The stark spreading skeletons of Pencil Pine (*Athrotaxis cupressoides*), King Billy Pine (*Athrotaxis selaginoides*, Plate 6.1) and Deciduous Beech (*Nothofagus gunnii*) are all too common features of a

Plate 6.1 Fire-killed King Billy Pine (*Athrotaxis selaginoides*) in the West Coast Range, Tasmania (A. Moscal).

high country degraded by the fires that were previously a rarity. Where there was a dense cover of alpine shrubs before a fire, twenty years after sees much bare ground and little shrub cover. The mountain peats record charcoal particles at the same depth as pollen changes reveal the local extinction of fire-susceptible shrubs, an extinction that lasts in some cases for thousands of years. Yet the rate of destruction of the Dwarf Pine and Deciduous Beech communities by fire cannot have been higher than in the period 1960 to 1980. Ironically, the rate of destruction of these communities might never be as high again because relatively little remains and the remnants are often protected by the persistent firebreaks created by previous fires or are found in naturally protected mountains.

In the alpine vegetation of the mainland, the only surviving community susceptible to long-term destruction by one or two fires is heath dominated by Mountain Plum Pine (*Podocarpus lawrencei*). This spreading shrub typically occupies extremely rocky areas where fire carries with difficulty and where the base of the shrubs is unlikely to be killed even if the tops are destroyed by radiated heat. The Mountain Plum Pine also has the advantage of a fruit that is dispersed by birds, enabling a more rapid recolonisation than most of the pines that are restricted to Tasmania. The Mountain Plum Pine is also the only pine that survives in the much burned alpine vegetation of Mt. Wellington, near Hobart, where previously the Creeping Pine (*Microcachrys tetragona*) also occurred. The Creeping Pine and Mountain Plum Pine are dispersed in the regurgitate of montane currawongs, but the other four alpine pines have no adaptations for medium or long-distance dispersal.

One of the most interesting of the fire-susceptible shrubs of Australia is *Microstrobus fitzgeraldii*, a species confined to the immediate vicinity of waterfalls in the fire-prone sandstone country of the Blue Mountains, west of Sydney. It is known from only six localities where 203 plants have been counted. The waterfalls protect it from fire, but the waters of some are now bringing a potentially lethal load of pollutants from the urbanised plateau.

Fire Management in Rainforest and Alpine Vegetation

The problem of the preservation of vegetation that is easily destroyed by fire is compounded by its incorporation in a matrix of vegetation that is not only easy to ignite, but burns and propagates fire with gusto. Tropical rainforest is often directly juxtaposed to highly inflammable grassland or grassy forest dominated by eucalypts. The temperate rainforest and coniferous and deciduous heaths of Tasmania are often only separated from the Button-grass (*Gymnoschoenus sphaerocephalus*) plains, which consist of highly flammable sedge and shrub species, by narrow bands of intermediate vegetation. On most days of the year such a juxtaposition is not a problem because, even if the adjacent Button-grass plains are alight, the other types of vegetation are too wet to burn.

Species that require fire for their regeneration, or their survival in competition with other species, generally have evolved characteristics that make them flammable to the degree that is necessary to ensure their success. Thus, the species that inhabit the most frequently burned areas tend to be more flammable than those that inhabit less frequently burned areas, while these in turn are generally more flammable than those that occupy areas that are seldom, if ever, burned. The species in the oft-ignited areas tend to have well-aerated foliage that is low in ash content and well endowed with oils that explosively ignite at quite low temperatures. These volatile oils are generally absent from the closely-packed foliage of species in infrequently burned areas. These species have high ash or water contents in their leaves, ash being a major ingredient in fire-suppressive chemicals.

Eucalypts propagate fire. The stringybark species send a shower of fine sparks from their burning bark up to fifty metres ahead of the main fire front. Gum-barked species have long strips of bark that dangle from branches and which can be carried, with other material, up to eighteen kilometres ahead of the main front. Thus, any bit of rainforest, coniferous heath or deciduous heath that happens to have its peaty soil dry out enough to burn on a day in which eucalypt forests are blazing has a good chance of igniting from the fall of burning embers. Such spotting can occur even if the burnable patch is well-surrounded by similar vegetation on soils not dry enough to burn.

The likelihood of spotting occurring in rainforest or alpine communities increases where such communities are intimately intermixed with more flammable vegetation, as the number of spots declines with distance from the source. The blowup fires that cause extensive spotting on days on which parts of the rainforest are dry tend to be associated with particular meteorological conditions. In Tasmania, the recent destruction of rainforest has been most extensive to the south-east of the mining settlements and main roads, as fire weather is north-westerly.

The rules for minimising future loss of rainforest and alpine vegetation from fire devolve into preventing ignition occurring to the windward during the short period of the year when these communities are dry enough to burn. This effectively means either keeping people away from the critical places or making sure that the people that are in them are motivated not to light fires.

Fuel reduction burning

Although little effort has been expended in trying to keep fire out of our remaining rainforest, a lot of thought and effort has gone into preventing fire from gaining access to pine plantations, tall eucalypt forest and built up areas. The major approach currently favoured by fire authorities has invariably been the reduction of 'hazard' through the agency of fire. This is variously called fuel reduction burning, hazard reduction burning, control burning or biomass reduction burning.

There is a sound rationale behind the apparently quixotic chorus of 'burn to stop bushfire'. Fires require a source of ignition — fuel — and wind speeds and dryness that will allow flames to spread. There is little that can be done about weather conditions — even their prediction has proved somewhat difficult in the temperate part of the continent. Arsonists are also uncontrollable. Therefore, the only way to control wildfire is to reduce the fuel.

This tactic is extremely effective when applied to a limited goal. There is no doubt that long grass and eucalypts mix poorly with houses in Australia. If gutters are kept clear and fire screens placed over any possible ingress of sparks into the dry interior of the roof or the underfloor space, and if a house is well-surrounded by fire-suppressive shrubs, green grass, gravel or concrete, the chances of a fire insurance claim are drastically reduced. Similarly the undergrowth and litter in some forests can be kept at a low level by burns that do not directly hurt the trees, if such burns are frequent enough and if the right weather conditions can be picked. Such fuel reduction burning is undoubtedly effective in preventing the loss of commercial timber.

The logical leap that has caused much unnecessary change and damage to the Australian bush is that if fuel reduction works to reduce the dangers of wildfire for houses and commercial trees, it is a good thing that will reduce fire hazard everywhere. Thus, throughout much of Australia teams of volunteers and government employees can be seen with torches, fire guns, lasers and fire bombing planes; all having great fun burning the bush. Hazard reduction burning has become an Australian spring rite.

The frequent planned burning of treeless areas has become a common practice in Australia. The value of this practice is often dubious. The burns are supposed to act as large firebreaks, which are believed to be valuable in controlling the spread of wildfire. The treeless areas also often have the virtue that it is possible to burn them without hurting anything 'valuable', with a bit of luck with the weather, whereas many commercial forests are either just regenerated and therefore too short to burn without damaging the young trees, or have such heavy understories and fuel loads that they are impossible to burn without damaging the timber resource. These latter types of forest will usually only burn readily on the occasional blowup day. Firebreaks are useful on blowup days as refuges for fire fighters, but rarely prevent fire spread in severe conditions because of the spotting that occurs from the eucalypts. Forest types that can burn on other than blowup days do not need the firebreak in less severe conditions, when all that a fire would do is reduce fuel levels. Refuges for fire fighters are perhaps best provided specifically, as are lines from which fire fighters can work to extinguish a relatively small fire that could present a larger danger. Also, most frequently burned treeless areas in forests will carry fire on extreme days soon after they have been hazard reduced, as the result of their possession of a fire-adapted species complement.

An increase in the frequency of burning of any area in Australia is likely to lead to the development of a vegetation type more encouraging of fire. Thus, relatively non-flammable broad-leaved shrubs will be replaced by the easily burned bracken or grass with an increase in fire frequency in many Australian suburban eucalypt forests. Near houses and under commercial trees the increased number of days on which ignition and spread of fire is possible will be more than compensated for by the reduction of fuel loads, and thereby damage, when a severe fire occurs. However, in other places this substitution may increase the danger of a severe fire by increasing the likelihood of successful ignition, and thereby increasing the probability that a fire will be available to initiate a conflagration when conflagration weather occurs. Of course, if all forest understories are reduced to bracken and grass or concrete or dirt, there can be no conflagrations. However, this total conversion would be so expensive, and so dangerous, that it is highly unlikely. As some of the largest reserved areas of the poorly-reserved tall wet forest ecosystems occur close to our major cities, the conversion would also be undesirable for conservation and aesthetic reasons. Far better to protect individual houses or to not build them in places of extreme hazard.

Broadacre burning tends to be carried out to a plan. The same areas are regularly burned at the same time of the year, and the burns are relatively cool and patchy. Cool burns do not encourage regeneration from seed, thereby leading to the gradual reduction of those species that do not propagate vegetatively. Some bush peas and wattles may be particularly susceptible to elimination through repeated cool burns. These are the main species that fix nitrogen in Australian soils. Cool burns consume the understorey while failing to scorch the tall shrub and tree layers. The taller plants are thereby able to grab more root space, and ground cover can be drastically reduced. The reduction of ground cover often leads to accelerated soil erosion.

Given that broadacre controlled burning is of dubious use in preventing the loss of property, the rationale for such activity should be ecological. In some circumstances, the absence of fire may endanger biological diversity. Planned burning may be the solution.

SEASONALITY OF BURNS

The season of burning can be critical in deciding the composition of bush. Some species will regenerate vegetatively after fire only in the seasons in which their carbohydrate and moisture reserves are high.

Species that regenerate from seed after fire breaks its dormancy may establish better in one season than another. Some annual species can be eliminated from a site by fire in particular seasons.

One of the best examples of this latter phenomenon occurs in the understorey of the open eucalypt forests in the monsoonal area of northern Australia. Although the exact timing of the wet season varies, it always comes and always brings with it a large amount of moisture. The beginning of the wet season in the country near Darwin sees a flush of germination of the tall annual *Sorghum* grass species in the understorey of the communities dominated by eucalypts. Seed is set and dropped during the wet season. By the start of the dry, a stroll through a eucalypt forest results in the accumulation of a substantial mass of straw around the ankles. In the meantime the *Sorghum* seeds have twisted themselves into the ground through the agency of a tail (awn) that corkscrews as it dries. Safely hidden in the soil, the seeds survive the fires that sweep the plains during the dry. If the firing is early in the dry some of the seed will germinate, heat being responsible for breaking a short-lived dormancy that normally prevents germination until the first rains of the next wet. These seedlings mostly survive to produce new seed during the next wet. It is fires lit during the early part of the wet that are capable of eliminating *Sorghum* from an area, because of the total lack of any long-term dormancy in the seed store.

The managers of Kakadu National Park are using such fires to create *Sorghum*-free breaks between compartments of vegetation burned, as is now normal, during the dry season. They wish to create such breaks because they believe the huge, late dry season fires that now sweep the park are less desirable than the patchwork quilt of smaller fires, lit earlier in the dry season, that resulted from Aboriginal land management.

THE INTERACTION BETWEEN FIRE AND GRAZING

Young plants and young shoots tend to be the most susceptible to animal grazing. They are rich in nutrients and poor in protection. Thus, species that might not be susceptible to elimination in normal circumstances may be grazed out after a fire. This is most critical for those species with no vegetative recovery mechanisms. However, even vegetatively recovering species may be ultimately reduced in numbers by intensive grazing of their reproductive organs, which often tend to proliferate after fire, an adaptation related to regeneration opportunities.

Much of the remaining Australian bush is grazed by cattle or sheep. Graziers make wide use of fire as a tool to provide better feed for their stock. Unfortunately, their perceptions of short-term benefit tend to cloud any consideration of possible long-term deterioration, especially during drought years when irreparable damage can easily result from economic desperation. The synergistic interaction of fire and grazing becomes most critical in extreme environments, and has created most concern in the alps and the desert.

There has been widespread degradation of desert and semidesert areas (see Figure 4.1). Here, dramatically different vegetation types can be produced by different combinations of firing and grazing. Much of the desert and semidesert grazing land of Australia has an understorey of hummock grasses, commonly known as spinifex (*Triodia* and *Plechtrachne* spp.). These grasses trap the blowing desert sand, thereby becoming hummocks, and often, as they grow, form rings. They protect themselves against herbivores by resinous deposits on their leaves and the high degree of pungency of their leaf tips. However, the young spinifex plants and the fresh second year shoots of spinifex form a major part of the diet of sheep. In the Pilbara region of north-western Australia, experimental work has shown that the maximum short-term productivity of spinifex was gained through the combination of summer burning and sheep grazing. Winter fires were patchy and encouraged the growth of unpalatable herbs and thickets of wattles (*Acacia* spp.), especially when combined with grazing. Areas converted to wattle thickets through this type of regime could be changed to a grassland of spinifex and other palatable perennial grasses by burning in summer once every five years. Of course neither firing nor grazing regimes are natural, so the vegetation is almost certainly different from that which greeted the first of the European graziers. The long-term consequences on soil and vegetation of the new summer burning regime are distinctly uncertain.

The consequences of the combination of burning and grazing upon alpine and subalpine grasslands and herbfields is a matter of far greater certainty. It is only in New South Wales that these practices have largely ceased, grazing continuing in the high country of Victoria and grazing and burning in the high country of Tasmania. In the Kosciusko high country of New South Wales, burning off largely ceased in 1951 and sheep and cattle were largely removed by 1961. Scientists in the CSIRO monitored the changes that occurred from 1958 to 1978, thus providing valuable information on the effect of the previous grazing of cattle and sheep and frequent autumn burning. Other information is available from long-term grazing exclusion trials in the Bogong High plains in Victoria.

In the subalpine zone, the Snow Gums had been eliminated by fire and grazing over some quite large areas. Fire alone might lead to the death of a large proportion of old Snow Gums in a stand, but regeneration would be prolific. However, grazing wears out the suppressed seedlings present from previous fires and eliminates any new seedlings, thereby ensuring the eventual death of the woodland. The Pencil Pines (*Athrotaxis cupressoides*) on the Central Plateau of Tasmania have met a similar fate, with each fire killing more of them, and with almost all regeneration being eaten by introduced animals.

Since burning and grazing ceased the Snow Gums have only recolonised sites that lacked a dense grass cover and were within the dispersal range of surviving trees, leaving vegetation patterns that relate more to paddock boundaries than to environmental gradients.

In the lower altitude subalpine valley grasslands of Kosciusko, the removal of stock grazing saw a shift in dominance from snow grass (*Poa* spp.) to Kangaroo Grass (*Themeda triandra*). The high altitude subalpine grasslands that had managed to maintain a high tussock cover, despite stock grazing and firing, experienced marked increases in the abundance of the more palatable herbs as these influences were removed. Where small intertussock spaces, sparsely covered with grazing-resistant herbs, were found, they were invaded by shrubs after grazing ceased. By about fifty years from the removal of grazing and the cessation of firing the shrubs will die and be replaced by grasses and large herbs.

Large intertussock spaces remain bare for at least twenty years after fire and stock are eliminated. In both the subalpine and alpine grasslands and herbfields, the bare ground created by grazing and burning is resistant to recolonisation because of the disrupting effects of needle ice. Needle ice is well-named, and occurs when below freezing temperatures are combined with wet soils directly exposed to the sky. The needles of ice lift the larger soil particles and plant seedlings out of the soil. Needle ice does not form under snow, where temperatures are an equable zero. Plant cover also prevents the formation of needle ice, as the foliage of the plants bounces back the heat that would otherwise radiate to outer space. Thus, large bare spaces can only be colonised through the lateral spread of plants established at their margins. This process is as slow as all growth in the cold of the highlands.

On deep, rock-free soils, the shrubs of the alpine and treeless subalpine zones are the equivalent of the secondary species of the tropical rainforest, forming the scabs that will fall to reveal the clean skin beneath. However, the shrubs are the bane of the life of the high country grazier. Stock do not like the highly aromatic Kerosene bushes (*Helichrysum hookeri*), or most of the other shrubs that invade small areas of bare ground. Therefore shrub growth is encouraged as the stock preferentially graze the grasses and herbs. The grazier sees his pasture visibly filling with uselessness — so he fires the pastures. This action appears to work as the snow grass puts up immediate new shoots. However, many of the shrubs resprout from rootstocks and the others take full advantage of the new bare patches created by the fire and subsequent stock grazing. The grazier in improving his short-term interests is virtually guaranteeing the medium-term destruction of the grazing resource. Such destruction is often effectively permanent, with all the fine soil particles eventually being removed by water and wind, leaving only a rocky pediment. Tussock grasses perched on pedestals of soil in rocky deserts are a common sight in the Central Plateau of Tasmania where much of the high country is privately owned.

The continuation of grazing and burning within the limited area of Australia's high country demonstrates political favouring of private interest over public good. The returns from grazing are infinitesimal compared to the returns that could be gained from the waters held within the soil and vegetation of the high country. Yet grazing drastically lowers the catchment value of the high country, and firing in combination with grazing speeds up the rate of loss of this value.

The high country vegetation, when undisturbed, protects soils from erosion by ice, wind and rain. Some vegetation types actively build up soil, thereby increasing the total possible high country soil storage of moisture. *Sphagnum* bogs are particularly good at both building up new soil and in holding moisture within the living vegetation. Unfortunately, communities containing *Sphagnum* have been among those most affected by grazing practices. They have also exhibited the slowest rates of recovery after removal of stock.

The importance of holding moisture in soil and vegetation relates to the economics of water usage. Vegetation uses water in photosynthesis, so might be thought undesirable in a catchment, and stock grazing therefore a lucrative way of increasing catchment values. However, in the absence of vegetation and soil, the water all runs off as soon as it falls, creating huge peak flows. Enormous dams would be necessary to capture these peak flows for use, and these would involve enormous unnecessary expense. If the peak flows were not captured for use they would become floods and create devastation in the lowlands. In between floods there would be little or no water in the rivers, creating further devastation. The filtering of water through the high mountain ecosystems softens the peaks and troughs of precipitation, mitigating flood and releasing water during drought.

The high mountain shrubs and trees have another important role, one not adoptable by post-grazing stone pavements. The fine particles of water that compose the mists that frequently envelop the high country do not fall as rain, but collect on leaves and branches and then drip to the ground. Thus, precipitation can be substantially increased by shrubs and trees. These shrubs and trees are the very same as those that the graziers regard as their bane.

INTERACTION OF FIRE WITH EXOTIC PLANTS

Just as the grazing of stock and fire can interact to cause vegetation changes that would not occur with either factor in isolation, fire is interacting with many of the overseas plant species that have found parts of Australia to their liking, to exacerbate their invasionary effects. Many of the most troublesome invaders of the bush, like gorse, broom and South African boneseed have seeds that store in the soil and germinate prolifically after fire. Thus, bush formerly lacking these exotics may be choked with them after a fire reveals the size of their soil seed store. Wind-dispersed exotics also take full advantage of fire-bared and fire-fertilised ground.

FIRE MANAGEMENT IN PROTECTED AREAS

Fire is one of the major problems of management in those parts of Australia that are devoted to nature conservation. The past is littered by appalling instances of management actions designed 'to combat the fire problem', most predicated on the usually erroneous assumption that the less fire the better. The firebreak was particularly favoured in the 1950s. Flinders Chase National Park on Kangaroo Island off the south coast of Australia has had its beautiful mallee cut into geometrical sections by firebreaks that are the envy of the farmers of the island and the despair of anyone with any feelings for the aesthetics of the bush. The Stirling Range National Park in Western Australia and Wilsons Promontory National Park in Victoria have also had firebreaks and fire roads inflicted upon their previously unflawed landscapes. In these places the cost has been greater, because the bulldozers creating the roads and breaks carried Cinnamon Fungus (*Phytophthora cinnamomi*), the most destructive pathogen ever to invade the Australian bush.

Broadacre hazard reduction burning is now more highly favoured than fire prevention. Where such burning is repeated at constant intervals and seasons and in a variety of vegetation types, it is likely to be a threat to the survival of some species and communities, as different species within the one community, and different communities, are adapted to different fire regimes.

The formulation of useful fire management plans requires a readiness to define the goals of management and a willingness to test the fruits of the plan against the goals. There are few vegetation types of which we have sufficient knowledge to be able to formulate a perfect fire policy for their perpetuation. Most of the bush, inside and outside reserves, looks the way it does and is composed of the species that it is, as the result of a highly varied pattern of fire, and other land-use practices, being imposed upon the results of the Aboriginal fire regime. The vegetation types that can be destroyed by fire, yet perpetuate themselves without it, present the least problem. Fire suppression is the only possible policy. All other vegetation types present considerable difficulty, this difficulty increasing as our knowledge of Aboriginal land management becomes foggier. Thus, a return to Aboriginal land management practices is more of a viable option in a place like Kakadu, where traditional knowledge survives, than in the South West National Park of Tasmania, where Aboriginal burning practices can only be poorly deduced from limited historical evidence and the nature of the vegetation.

Even if it were possible to discover the Aboriginal fire regime for the areas now contained within National Parks and Nature Reserves, it might not always be either possible or desirable to repeat it. The effects of fire in interaction with the exotic plants and animals now present in most reserves might be quite different from the effects of the same fire in a natural situation. Given the small areas within most reserves, it might be impossible to maintain populations of rare plants and animals while applying the Aboriginal fire regime. In some cases, the Aboriginal fire regime might present an unacceptable hazard to the neighbours of a reserve. In other cases, the pattern and nature of the vegetation might have changed so much since European settlement that the results of the imposition of the old fire regime could be perceived as disastrous.

One of the major goals of a National Parks system should be to maintain the maximum possible variety of species and communities. This goal almost certainly requires an enormous variety of fire regimes, the variety occurring in both time and space. Knowledge is needed of the fire regime boundaries within which particular species and communities will survive in the long term. This knowledge can be gained through careful observation and experimentation. Long-term observations by park managers of the biological results of both accidental and deliberate fires, and basic data on the life history and fire adaptations of organisms, could provide a firmer base for fire management decisions than is available today.

Where there is a high degree of uncertainty about the appropriate fire regime for the perpetuation of the species and communities within a

National Park, small-scale experimentation with different reasonable alternatives within vegetation types, and the general maintenance of a variety of fire regimes, including no fire, would take the least risk with the future of our bush and its component species.

■

7

The invaders

As the continents drifted apart so did the evolutionary pathways of the assemblages of species that occur within similar but well-separated environments. Ancient barriers of ocean, heat and cold divided the world into biotic realms which have few native species in common. Thus the rainforest of the Americas differs markedly in its species composition from the rainforests of Africa and Southeast Asia, which themselves have few common species. Yet the plants growing wild in ground disturbed for farm and city differ little between Cairns and Manaus on the Amazon. The explosion of European people over the surface of the globe has seen an unprecedented mingling of previously discrete floras and faunas, a mingling that is still taking place.

Introduced, exotic and alien are adjectives used to describe species growing outside their natural range. In Australia the natural range of a species is taken to be that immediately preceding the European invasion. Most introduced species are considered to be weeds, which are defined as plants we do not want, or plants out of place. However, natives such as Austral Bracken (*Pteridium esculentum*) are often considered to be weeds, and many people regard many introduced plants as valuable. Environmental weeds are those exotics that penetrate the bush.

A BRIEF HISTORY OF INTRODUCTIONS

Australia was a late recipient of the seeds of the world. Despite this late start, approximately one in ten plant species growing wild in Australia has been introduced from another continent since the European settlement of Sydney Cove in 1788. The number of naturalised introduced species has increased almost linearly since the European invasion. Because their number is constantly increasing and because we are not quite certain whether many species are introduced or native, or whether one species, as with the Blackberry (*Rubus fruticosus*) and Soursob (*Oxalis corniculata*), is actually several, no exact figure can be given. Even the number of native higher plant species can only be safely guessed to be between 15,000 and 25,000, the difficulties and disagreements of plant taxonomy being what they are.

In the early decades of settlement, horror at the untidy greyness of the Australian bush impelled a rapid introduction of useful plants and tidy, green and floriferous ornamentals. Gorse (*Ulex europaeus*) was introduced as a hedge plant.

Gorse is still listed as a garden plant in one of the most popular compendiums of ornamentals, along with other virulent bush invaders such as Himalayan Honeysuckle (*Leycesteria formosa*) and Lantana (*Lantana camara*). Half of the introduced species found in the Victorian bush in 1989 were available from commercial nurseries. Blackberries were disseminated throughout Victoria by the famous botanist Baron von Mueller to provide sustenance for travellers.

Although deliberately introduced plants have transformed much bush, more exotic species have gained our shores by accident than by design. The ceremonial spraying of the interiors of aircraft arriving from abroad may kill the stray emigratory gnat, but has no effect on the grass seeds lodged in a passenger's socks, or the fungal spore hidden in dirt stuck to a boot. As international trade and travel increase, so do the chances of successful immigration of these organisms.

In the early years of the nineteenth century there must have been swarms of accidental arrivals. The seed of unwanted plants would have been mixed with crop seed, feed for transported stock would have been full of the seed of potential new arrivals, and materials transported as ballast would often have been similarly full of seeds. Some idea of the role of stock in the spread of exotics can be gained by the relative proportions of aliens in the floras of two Bass Strait Islands, Rodondo and Hogan. The former island has been seldom visited by humans because it is surrounded by steep cliffs plunging into some of the roughest waters in the world. Only four exotics out of a total of 56 species were recorded in a survey of this island in the early 1970s. In contrast, nearby Hogan Island, which had been grazed for 80 years, had 58 exotics out of its total of 146 species.

Is disturbance necessary for exotic invasion?

The presence of exotics on a remote, hardly visited island such as Rodondo intimates that one of the most comforting axioms of Australian ecology might be an unfortunate myth. If human disturbance was necessary to allow the spread of exotics in natural vegetation, the management of exotics in natural bush would present few problems: no human disturbance — no exotics. Unfortunately, native animals and natural geomorphological processes are capable of creating disturbed ground that does not differ in its qualities from that created by people. Bush burned as a result of lightning strikes is just the same as bush lit by matches. Tomato seeds defaecated by people germinate no more effectively than those defaecated by native animals. Some introduced plants even have the temerity to make their own way into native bush that has not been dug up, burned or defaecated upon by either nature or humans. Some of these latter species are the major villains of the bush.

Characteristics and distribution

Most of the species introduced into Australia in the last 200 years are followers of human habitation. Like the louse, they would mourn our demise. The weeds that we find in our gardens, fields and vacant blocks have developed life histories that allow them to take advantage of the types of repetitive and frequent disturbance that characterises the activities of human beings. There are weeds whose seeds have developed the same specific gravities as the crops whose fields they share. There are those whose seeds drift widely in the wind, looking for the broken earth. There are those that use mice or starlings to spread their seed. There are those that multiply when torn apart by spade or plough. The soil is full of seed. In the bush it is mainly that of native plants. In the city and field it belongs mainly to the exotics. That so few of our weeds are natives testifies to the recency of agriculture in our biotic realm. That so many of our weeds are European testifies to the long history of agriculture on that continent and to the origins of our recent human immigrants. The *Illustrated British Flora* is still the best available book for the identification of most temperate Australian weeds.

Not all of the non-woody weeds of farm and city significantly penetrate the bush. In Victoria approximately half of the naturalised alien plant species are confined to cultural vegetation. Species such as Flatweed (*Hypochaeris radicata*) and Yorkshire Fog Grass (*Holcus lanatus*), the latter the dread of hay fever sufferers, have had some reasonable success, but do not threaten to overwhelm the native species, rather being a new element in the understorey mixture. Much of the remaining bush occupies an environment too hostile for plants accustomed to the rich soils of farm and city. Australian soils are generally very much poorer in nutrients than those found elsewhere in equivalent climatic zones. This poverty of nutrients has been attributed to the relatively ancient nature of the land surface. Most introduced species cannot cope with macronutrient and micronutrient deficiencies that prove no inhibition to the survival and success of a large proportion of the Australian native flora, just as many of the native species find it difficult to cope with normal agricultural levels of nutrients. Most of the bush on the richer Australian soils has been cleared, giving many exotics little scope for bush invasion.

In southern Australia, environmental weeds tend to occur together in highly dynamic communities consisting of a synthesis of native and exotic plants, although a few of the more threatening species, such as Gorse and Marram Grass (*Ammophila arenaria*) can form monospecific stands. In northern Australia monospecific weed invasion is much more common than the synthetic community pattern, although both are present.

Time and distance are critical variables controlling the spread of those exotics that depend on the disturbance of the soil for their establishment. Constantly disturbed sites in remote areas may lack exotics because they have been hitherto unable to cross the intervening undisturbed bush. Yet a few seeds very rarely might be carried the requisite distance. Given enough time one of these seeds will germinate on the disturbed site, then grow into an adult producing more seed. If the progeny of this plant are as successful as their parent the native flora of disturbed ground will rapidly be displaced by an introduced flora of disturbed ground. The probability of such a migration taking place increases as the number of plants of any exotic species increases at a constant distance from the location of the potential colony, and also increases with time and proximity of exotic plant populations. The same rules apply to the spread of those exotics that do not require repeated disturbance, except that the barriers to their spread are usually fewer.

EUTROPHICATION AND INVASION

The bush in close proximity to suburbia has its soils enriched by septic tank outputs, the dumping of rubbish, airborne particulates and household drainage. In the Sydney region there is on average 50 parts per million more phosphorus in bush soils enmeshed in suburbia than in those remote from suburbia, a substantial proportionate increase.

The phenomenon of cultural enrichment, or eutrophication, of soils has been used widely by archaeologists to locate old settlements, as they can be detected by the lushness of plant growth, even if not by surface features.

The effects of increasing nutrients to a level suitable for the growth of agricultural plants has been demonstrated in a series of experiments in heath, the vegetation type characteristic of some of the poorest soils in a continent characterised by poor soils. Both in the Wallum wet heaths near Brisbane and in the dry heath of the Ninety-Mile Plain in South Australia, the effects of the fertilisation of soils were to temporarily encourage the growth of some native species, poison others, allow the invasion of exotics, and cause an ultimate almost total transformation in the nature of the vegetation.

In the Wallum, as a result of the faster growth rate allowed by the accession of nutrients, a native tree species grew above the height at which it ceased to be set back to ground level by fire. Most of the original plant species disappeared from the fertilised area, which ended up with an understorey of grasses and introduced shrubs. In the Ninety-Mile Plain heaths, the native shrub species could not establish in competition with the exotic herbs after fire. Exotic weed dominance seems to follow fertilisation of the nutrient-poor ecosystems (characterised by hard and small-leaved shrubs) to a greater extent than in the ecosystems with naturally higher nutrient levels (characterised by grasses and/or broad-leaved shrubs), although the latter ecosystems are more susceptible to minor invasions of exotic species.

Any bush near farm land is likely to collect some nutrients, aerial topdressing of pastures being a normal rural activity. Nutrients can also build up from the most unexpected sources. Only 1.1–1.4 days of normal individual human urine output will fertilise a square metre of soil sufficiently to raise it from being able to support only native heath to levels suitable for crop plants and most weeds. Eighty-three banana skins will do the same job. The equivalent figures for some other commonly deposited substances are: steer faeces, 0.2–0.4 days worth; dog urine, 2.7–4.0 days; cat urine, 4.0 days; orange peel, 37; apple cores, 114; cigarette butts, 375. Thus, the weeds along tracks in national parks can be attributed not only to transportation of their seeds by people, but also the transformation of the soils by their effluvia. Popular camping places in the highly nutrient-poor walking country of Tasmania have an urination ring of introduced annual plants. This growth is most prolific directly outside the doors of huts, a comment on Tasmanian weather as well as male night laziness.

The inevitable raising of nutrient levels in small areas of bush surrounded by development means that there is no way, short of topsoil scraping or cultivation, to perpetuate the types of vegetation adapted to nutrient-poor soils. Even cultivation could fail for some species if nutrient levels became too high. However, there may be means of maintaining native bush on better soils despite a close proximity to development and the invasion of potentially transforming introduced plants.

TIME BOMBS

Fortunately, the species that are capable of insinuating themselves into the bush in the absence of repeated disturbance tend to disperse for lesser distances than those that depend on disturbance. Or perhaps this is not so fortunate, for changes that take place over a long time are less readily recognised as dangers than ones that take place with devastating rapidity. The spread of willows (*Salix* spp.) along streamsides has been both rapid and widespread, to the extent that they are regarded as a fixed part of many rural and bush scenes. Yet species with a short normal dispersal range, such as the Californian Redwood (*Sequoia sempervirens*) and the Monterey Cypress (*Cupressus macrocarpa*), might have the potential to invade much larger areas of bush than the willow. Both species have been planted widely through regions of similar climate and soils to those in their homeland. Seedlings of both species have established of their own accord near planted parent trees. Californian Redwood is as fire-dependent for its regeneration as the native Mountain Ash (*Eucalyptus regnans*) (see Chapter 5), but lives over 2000 years compared to the 400 years of the Mountain Ash. The redwood is also more fire-resistant than the Mountain Ash, being capable of recovering vegetatively from fire to a greater degree than the eucalypt, and creating such a dense canopy that the probability of the spread of fire is much lower than can ever be the case for a eucalypt forest. Monterey Cypress has the same ecological characteristics as the native cypress pines (*Callitris* spp.), and may have the potential to replace them in some parts of their range. It certainly has the potential to displace much coastal eucalypt forest in a situation of low fire frequency.

We probably have many species well-established in cultivation that are yet to exhibit their invasive qualities. Some of our most virulent invaders have been present in Australia for many decades before they exploded into the natural landscape. *Mimosa pigra*, a prickly tropical shrub from Africa that is threatening the future of the native vegetation of the flood plains of northern Australia, took a century to break away from the gardens of Darwin. Such protracted periods of quiescence can be related to the time taken to develop genotypes suited to local bush conditions, the probability of escape from cultivation and the initial slow absolute increase in the exponential population curve that is typical of new invaders.

SUBURBAN AND FARM ESCAPEES

A large proportion of the introduced shrubs that we plant in our gardens have already started to spread elsewhere. Pome fruit trees are present on New England roadsides in the same proportions as their sales in the Brisbane market, and they can also be observed established along walking tracks within National Parks. Those ornamentals with fleshy fruits that are suitable for dispersal by birds have proved particularly virulent invaders of the bush. Privets (*Ligustrum* spp.) have filled many a Sydney gully, disseminating from the neat yellow hedges of suburbia. Cotoneasters (*Cotoneaster* spp.) and Hawthorn (*Crataegus monogyna*) invade the wet gullies near Hobart, where Hawthorn is in turn invaded by the native gully trees that take advantage of its deciduous habit. Boneseed (*Chrysanthemoides monilifera* ssp. *monilifera*) has overwhelmed many remnants of native bush in the Melbourne Metropolitan area. Olives (*Olea europaea* ssp. *europaea*) cover the hills near Adelaide. All the above invasions seem to fit the disturbance axiom, in that they are associated with roads, tracksides and the nutrient-rich interface of suburbia and bush.

Most of the shrub species that so vigorously invade urban bush are quite capable of invading unenriched bush elsewhere if the opportunities of dispersal and a suitable environment permit. An irrefutable example of this ability is provided by a native species that has burst out of its original limits since being adopted as a garden plant.

The Sweet Pittosporum (*Pittosporum undulatum*) has a pleasantly dense, green, spreading canopy, beautifully scented cream flowers and attractive orange fruits. Before its adoption as an ornamental it was found solely as an inhabitant of rainforest and rainforest margins in south-eastern Queensland, coastal New South Wales and eastern Victoria. Now it has extended its range throughout the coastal regions of Victoria and Tasmania, as well as to Jamaica, Hawaii, Bermuda, the Canary Islands and New Zealand. Gardens have been the foci of its rapid spread, while the introduced Blackbird has been the major agent of its dispersal. Consequently it tends to begin its invasion of bush under the trees in which the Blackbirds perch and defaecate. The establishment of the species depends on freedom from fire, its thin bark making the Sweet Pittosporum highly susceptible to this element. Once established on a site it is not easily dislodged, as it creates closed canopy conditions that discourage fire and prevent the successful regeneration of eucalypts. It can establish as a seedling in its own shade. Thus, there is no obstacle to its continued spread apart from a high-intensity fire.

Control of environmental weeds

Just as Sweet Pittosporum could be eliminated in the early stages of its invasion by manipulating the environment using fire, many other species that have invaded the bush have weaknesses that can be exploited in order to achieve their local extinction. For example, the establishment of a closed-canopy forest will eliminate Gorse, although its seeds will linger in the soil. Stock grazing, or very frequent fire will eliminate Boneseed, although care would need to be taken to ensure that the cure was not worse than the disease for the bush in general. Unfortunately, the smorgasbord of aliens present in a lot of bush often makes a cure for one an encouragement for another, in the same manner that pesticides often create new pests.

The form and intensity of management of the exotic plant species that are invading (and will continue to invade) the Australian bush will be critical in deciding its future. Most of the detailed knowledge of the biology and ecology of the bush invaders we possess because they are also nuisances in farm land or because difficulties were experienced in utilising a species for some practical purpose. For example, the germination requirements of Boneseed were researched because of difficulties encountered when using the species as a stabiliser of dunes reconstituted after sand mining, and the biology of Gorse and Blackberry are relatively well known because of their impact on pastures.

The contemporary agricultural form of attack on unwanted plants is to hit them with herbicides, usually lifeform-specific growth hormones. They then obligingly grow themselves to death while the plants that are really wanted sit unharmed. This high-technology approach to weed control cannot be satisfactorily adopted for most conservation purposes because the target organisms are usually mixed in with equally susceptible organisms that are to be protected. However, it may be possible to use herbicides directly on environmental weeds by wick application, cut-stump application or directed spraying.

Dense infestations of weeds in native bush need more than spraying. The weeds may be killed but their seeds stored in the soil will usually replace them, and if the original weed species do not survive others will emerge phoenix-like to take their place.

Cultivation is probably the only solution for the densest weed infestations. Thickets of privet can be bulldozed, then replaced with a regularly mown lawn consisting of native grasses, pleasantly planted with other species that originally inhabited the site. If the dense infestations are left to fester, seeds will continue pouring from them into the adjacent less-disturbed bush. However, the critical front for the repulsion of introduced weeds is where their growth is thinnest. Small introduced plants and seedlings can be mercilessly plucked from the ground, and all divots replaced and carefully covered with native plant litter. This Bradley method makes human beings a selective force in favour of the Australian native plant species, and is now gaining widespread favour in the management of small bush parks. The method is specially effective for the control of woody weeds. Well-dispersed and rapidly proliferating introduced herbs and grasses may be less tractable to manual control. For example, the removal of even light infestations of Common Oniongrass (*Romulea rosea*) would involve digging out the bulbs, in the process of which the soil compaction with which onion-grass invasion is associated might very well be increased. Some invaders are so widespread that their elimination is effectively impossible, and most of these species do not seriously displace natives. Thus, to avoid unrewarding labour, a discrimination needs to be made between those species that are easily and fruitfully eliminated and those that do little harm, create more harm in their elimination than in their persistence, or are impossible to eliminate.

The Bradley method, involving systematic and gradual clearing of weeds to allow the natural re-establishment of native vegetation, is so expensive in labour time that in effect it relies on volunteers. Thus, herbicides and other more labour-efficient methods are often used in areas lacking access to volunteers. In some circumstances mowing, slashing and chaining can often be used to deplete or eliminate populations of environmental weeds while allowing native species to survive.

There have been several instances where a virulent weed has been reduced to decrepitude by the deliberate introduction of organisms that look upon it as breakfast, lunch and dinner. One reason why so many introduced plants meet with so much success in a new land is that they are likely to migrate without their full complement of insects, mammals, fungi and microbes that make life a struggle in their native land. However, the introduction of pests is not without hazards. Unless the biological control agent is highly restricted in its dietary requirements, the intended victim may be abandoned in favour of juicier natives. In the famous case of the Prickly Pear (*Opuntia* spp.) and its nemesis, the Cactus Moth (*Cactoblastis* sp.), there was not much chance of greener pastures, succulence not being a favoured adaption in Australia. However, the rust that has been released to attack the Blackberry problem, may reduce the populations of the native Small-leaf Bramble (*Rubus parvifolius*), a more attractive and less virulent species than the invader.

Boneseed, the bane of the bayside bush, is particularly susceptible to biological control. It is a South African coastal daisy species, introduced to Australia for its ornamental qualities and dune reclamation. It is pretty enough when its foliage is obscured by yellow flowers, but the flowers are followed by bone-hard seeds, luringly encased in soft flesh. The seed happily survives passage through the digestive and excretory systems of birds, being consequently well dispersed and densest under trees used for roosting. Boneseed has been particularly successful in invading and displacing stands of Coast Wattle (*Acacia sophorae*). Stands of Boneseed produce more than 4000 seeds per square metre, while the wattle produces only 100 seeds which are, moreover, mostly lost to predators. Differences in the production, dispersal and predation of seeds lead to soil stores of the viable seed of the introduced plant amounting to sixty times that of the native wattle. However, in South Africa, where Australian wattles are invading communities containing Boneseed, the soil store of the viable seed of the wattles is fifty times greater than that of Boneseed. In South Africa the seed of Boneseed is eaten, but not that of the wattles. There is therefore some considerable potential for the location of seed predators that devote their days to attacking Boneseed, in order to achieve some control of this destructive weed without destroying its value as an ornamental.

Preventative action is better than curative action with environmental weeds. Import restrictions on plants are currently inadequate. Several of the top ten environmental weeds recognised by the CSIRO have been introduced as fodder plants and the remainder are ornamentals. Fodder plants and ornamentals are still being introduced to Australia, without any testing of their potential impact on our native vegetation. The minimisation of activities that aid the spread of exotics is also highly important. Vehicles have been shown to be excellent plant vectors. This is one of the good reasons to maintain wilderness, which by definition cannot contain roads. The local regulation of plants available for nursery sales is also critical, because species that are environmental weeds in one area may be benign ornamentals in another.

Our most threatening plant invader—the Cinnamon Fungus

The class of species that are impossible to eliminate includes an organism that is having a greater impact on the Australian bush than any other plant species. The Cinnamon Fungus (*Phytophthora cinnamomi*) was first discovered as an inhabitant of the bases of Cinnamon trees on the island of Sumatra. It next drew attention to itself by gaining a living rotting the roots of avocado trees, an activity that impelled avocado planting on to the well drained, steep slopes that inhibit the activity of the fungus. Sometime in the last 200 years the species made its way to the Australian bush. Some scientists argued that the Cinnamon Fungus is a native that only becomes a nuisance after human disturbance of the bush. However, undisturbed bush has proven no barrier to its spread, and if it is native, it is certainly not to the southern part of the continent.

The Cinnamon Fungus makes slow progress when unaided by people. Its spores can be carried in ground, water or overland flow, but across and upslope it will move only at the rate of a few metres a year. Over millennia it might be carried long distances in the dirt attached to animals or birds, but large amounts of fresh infested soil are necessary to give a good chance for its migration.

The fungus was most probably carried to Australia in soil in which imported plants were growing. It has certainly become well established in the garden suburbs of Australia. The dumping of garden waste would soon have insinuated the fungus into the bush. Two of the remotest gardens in Australia, those around the only residences on Bathurst Harbour in the far south-west of Tasmania, are riddled with the fungus, to which most well-fed domestic plants are reasonably resistant. They have acted as the foci of infestation by the fungus, which has spread along the walking tracks of the south-west.

The fungus has been relatively quickly transported along the mud wallows that pass as tracks in south-west Tasmania, but along most tracks in Australia there would be much less chance of transport by boot. Instead the fungus is transported by the largest and fiercest inhabitant of the Australian bush since the extinction of the megafauna. The bulldozer is the ideal vector. The justification for its existence is its ability to tear the surface of the earth, and in doing so it collects large amounts of soil in its many recesses, soil that can easily be dislodged in the churning of the next job.

Cinnamon Fungus was not recognised in the bush until well after it had begun to create enormous amounts of damage. Before the forest managers and workers became aware of the dangers of fungal pathogens in general, and this one in particular, a dying area of forest was regarded as ideal for a gravel pit or quarry. That way, there would be no need to destroy good forest land to get material for road building and maintenance. As a result, the fungus experienced an accelerated expansion along road verges, from which it was further disseminated as machinery moved into adjacent bush for logging operations.

Cinnamon Fungus has been recorded as a cause of native plant death from the rainforests of Queensland to the Button-grass plains of south-west Tasmania. The Jarrah forests and heaths of Western Australia, the Silvertop forests of East Gippsland and the heathy communities of south-eastern Australia are being most heavily attacked. When the soil is warm and moist the root systems of most plants can be attacked by the fungus. A plant will die when its root system is so eaten away that it cannot take up enough water during drought. Thus, the spread of dieback lags behind the spread of the fungus.

The presence of dieback caused by Cinnamon Fungus is generally marked by the high death rate of most shrubs and trees and the low death rate of most sedges and grasses. The pea shrubs (Fabaceae) and wattles (Mimosaceae) tend to be more resistant than the heaths (Epacridaceae) and banksias (*Banksia* spp.). Drought dieback can evince a similar appearance of brown dots among the green, but in the case of drought it is usually only the species at the dry extreme of their distributions that are affected.

The species that are most susceptible to death as a result of the root-rotting activities of the fungus are those that only make root growth in the soil temperature conditions in which the fungus is most active. Thus, although the fungus is found on the root systems of some eucalypts in the highlands of East Gippsland, its effect is minimal, the trees making up for damage when the fungus is quiescent. Similarly, the fungus cannot attack the root systems of the rainforest trees in south-west Tasmania because the dense rainforest canopy keeps temperatures constantly below those necessary for its activity. However, the soil of the Button-grass plains is exposed to the sun and develops high temperatures, so the fungus is active. The Jarrah has the misfortune to have its root growth totally confined within the period in which the fungus is active. Consequently, tall, healthy Jarrah trees are now confined to particularly well drained sites and those areas not yet penetrated by the fungus.

The fungus is less virulent on the soils of moderate to high nutrient status than on the low-nutrient laterites of the Jarrah belt and the impoverished deep sands of the heath country. There is some indication that, in some way, perhaps linked to their effects on the composition of the soil flora and fauna, nitrogen-fixing plants inhibit its activity. It may have been unfortunate that low intensity hazard reduction burning was adopted in the Jarrah forests at the same time as the rapid and unrecognised spread of Cinnamon Fungus. This fire regime favoured the understorey dominance of the heavily attacked banksias, and worked against the dominance of the nitrogen-fixing wattles, which perform better with an occasional severe fire. Unfortunately, a return to a wattle-favouring fire regime has not proved effective in reducing the impact of the fungus on Jarrah.

There seems little chance that a way will be found to destroy or control the Cinnamon Fungus without destroying the bush it attacks. Fungicides will kill it, and most other fungi, but the scale of infestation is such that control measures are impracticable. When the application of fungicides was tried at Wilsons Promontory National Park the fungus had already escaped downhill into a swamp. Attempts are also being made to identify organisms that live off the fungus, in the hope that these might be disseminated.

If biological control does not eventuate, much of the southern Australian bush is doomed to enormous change. Where stringybark or peppermint eucalypt species are mixed with gums, the former survive only as sickly seedlings and saplings wherever the fungus has made its way in lowland forests. Most of the attractively flowering shrubs that characterise heath and the heathy understorey of forests on sandy or gravelly ground are eliminated or almost eliminated, leaving drab sedges and rushes to provide most of the cover. The bizarrely beautiful grass-trees (*Xanthorrhoea* spp., Plate 7.1) are among the most susceptible types of plants. Some species, such as the prickly but floriferous Conebush (*Isopogon ceratophyllus*) may be rendered extinct by the fungus. Others, such as the Pink Swamp Heath (*Sprengelia incarnata*), may be eliminated from all but their alpine and subalpine moor habitats.

The speed of the spread of the fungus has not allowed the selection and adaptation within species that could have taken place with a slower progression unaided by man. At present, most of the susceptible bush has not been ravaged. However, Cinnamon Fungus is present in small and growing strongholds virtually everywhere. The only areas from which we could hope to exclude it in the long term are islands. This exclusion would require the careful cleaning of possible sources of contamination. Elsewhere its spread can be slowed by the adoption of such measures, but cannot be prevented.

Plate 7.1 *Xanthorrhoea*, one of the taxa most susceptible to *Phytophthora cinnamomi* (F. Bolt).

THE IMPACT OF INVADING ANIMALS

The Australian bush is paying for the lack of ecological understanding exhibited by the enthusiastic members of the nineteenth century acclimatisation societies. The members of these societies thought to increase the variety of fauna in their new land, little realising the degree to which the native flora and fauna would suffer from their activities.

All introduced grazing and browsing animals must inevitably cause some vegetation changes, because their food preferences will be different from those of the native grazers and browsers and because the native plants do not always possess adaptations that allow them to survive and reproduce in the face of the tactics adopted by the newcomers. Their preadaptation to the introduced herbivores gives introduced plants an advantage. For example, at Koonamore in South Australia a native speargrass (*Stipa* sp.) is replaced by the preadapted introduced grass, *Schismus barbatus*, where stock grazing takes place, but maintains its dominance where stock are excluded.

Rabbits are far from the only introduced animals capable of wreaking havoc in the bush, which contains domestic animals and wild populations of almost all domestic animals, up to and including the camel. Animals that root and wallow create particularly spectacular effects, including the dispersal of Cinnamon Fungus in the case of the feral pig.

Introduced and naturalised predators such as the cat and European Wasp may have largely unpredictable effects on the nature of the bush through their impact on the native animal species. Both herbivorous and carnivorous introduced species are likely to change the patterns of dissemination of both native and introduced plant species. The virtual elimination of medium-size native mammals within the range of the Red Fox has changed the nature of ground disturbance in many plant communities.

There are some introduced animals whose elimination from the Australian bush, or at least parts of it, is not beyond the bounds of probability. The wild Cape Buffaloes of Arnhem Land may be tamed within the next few decades, more because of the dangers they present to the cattle industry than because of their undoubted severe impact on the lagoons, flood plains and monsoon forests of the north. However, even the real danger of diseases spreading from wild buffalo to domestic cattle has not reconciled many of the inhabitants to the extermination of wild populations of a beast that they have come to regard with a regional affection and to use to generate regional income.

Cattle diseases have occasionally been an effective conservation argument in Australia. For example, the Queensland Government declared some large parks on the Cape York Peninsula because it thought that they would act as a barrier to the spread of disease from New Guinea.

Unfortunately for the bush, the few positive aspects of exotic animal invasion are far outweighed by the enormous changes they are inflicting upon the native Australian biota, and by the negligible chances of the elimination of most of the harmful species. The managers of land devoted to nature conservation will have to continue to devote much of their effort to reducing the numbers of the more transforming of the exotic animals, as leaving most parks to themselves will be a recipe for undesirable change. Some types of bush will only persist if particular introduced animals can be eliminated for a sufficient time period to allow the regeneration of the dominants. Others might require only persistent hunting at a level sufficient to maintain low populations of the crucial invaders. The importance of keeping those parts of Australia that are, by historical accident, free of particular animal pests cannot be overemphasised. If the fox, pig and goat can continue to be kept out of the Tasmanian bush, and the rabbit from the bush of Kangaroo and Flinders Islands, we will maintain types of bush that might not otherwise survive. Islands large enough to be rich in native animal and plant species but small enough that any invaders can be economically repelled, will have enormous importance in the future of many types of bush. Thus, their reservation is much more important than the gaining of similarly diverse areas on the mainland.

OUR MOST THREATENING ANIMAL INVADER — THE RABBIT

The rabbit is a generalist herbivore that excavates substantial burrow systems where the soil allows and occupies open environments from the alps to the desert. There are few large open parts of temperate Australia free from this species.

Mallee regeneration and dense rabbit populations appear to be incompatible. The mallees, multistemmed eucalypts with gigantic woody tubers that used to be much prized for firewood, dominate the natural vegetation of most country around the Great Australian Bight. Much mallee has been cleared for wheat cultivation and pastures, with variable economic success. In the drier part of its range, where it is underlain by salt bushes or hummock grasses, regeneration has probably always been a rare event, associated with heavy rain soon after fire. The only reason that the impact of the rabbit has not become more apparent is the ability of mallee eucalypts to put up new shoots from their underground woody lignotubers. These lignotubers can enlarge and rot in the middle while still being healthy on the outside, where a ring of apparently unrelated, but genetically identical individuals can be observed. Carbon-14 dating has shown the oldest of healthy wood to be 200 years, but some of the rings may have much older ancestry. Nevertheless, they are unlikely to be immortal, and the introduction of the rabbit could eventually result in the disappearance of much mallee.

Mallee is not the only vegetation type imperilled by the rabbit. The arid zone of Australia has already been grossly changed where sheep grazing has been practised, but even the exclusion of sheep may be insufficient to ensure the survival of many of the arid zone perennial plants. At Koonamore Station in the arid zone of South Australia part of a paddock was fenced off from stock in 1925 after 55 years of continuous grazing. The fencing was insufficient to totally exclude the rabbit, which in drought years can make a bridge of bodies over the stoutest fence. Drought even motivated the rabbits to climb wattle trees to five metres above the ground. Many died tangled in the nutritious branches. The only woody species that ceased to decline with the exclusion of stock were those with leaves too salty for the rabbit but not for stock. Even these species are not readily returning to the areas of scalded soil that resulted from generations of stock grazing. Many trees and shrubs, including Mulga (*Acacia aneura*), may only survive in the long term in the northern part of the arid zone, where the rabbit experiences climatic difficulties. Many other species are in the same boat. Experiments, during part of which *Acacia* seedlings were subjected to grazing by rabbits but not sheep, have shown that the rabbit can quickly find and destroy plants that constitute less than 0.00001% of the total available feed.

Controlled fencing experiments in the subalpine zone in which some plots were grazed only by rabbits, others by rabbits and native grazers, and others only by native grazers, verified empirical deductions about the large impact this introduced species has on high mountain vegetation.

The future of the rabbit in Australia is uncertain. The CSIRO is using genetic manipulation techniques to develop controls stronger than myxomatosis, but there is no guarantee of scientific success, which, if forthcoming, may be followed by political blockage.

Conclusion

After two centuries of European settlement there are few parts of the Australian bush that lack any introduced plants. Their area is constantly diminishing. There are probably no areas lacking introduced animals. All the published information suggests that the process of exotic invasion of the Australian bush is continuing apace, with the spread of already established species and the invasion of newly introduced environmental weeds. It is likely that a future, more stable Australian wild vegetation will consist of a large proportion of the extant native species mixed with multicultural assemblages of aliens. It is to be hoped that at least some of our larger reserves can be kept free of most exotics in perpetuity. Such a freedom will require constant vigilance.

■

8

Conserving the bush

The preceding chapters have established that the bush is disappearing at a steady and substantial rate. It is also undergoing more subtle transformations. If we wish to maintain ecosystems that consist of native species somewhere close to their number and relative abundance in the bush today then preventative management is needed. It is easiest to maintain biological diversity where it forms the over-riding management priority. National Parks, Nature Reserves and most other types of protected area usually have this priority in the sense that the major management goal is to avoid losing any native genotypes, species and communities in the reserve. Given that land clearance is effectively irreversible on the timescale of centuries to millennia, there is a strong and well-recognised need for protected area systems that will be secure and viable in the long term and that include all those elements of biodiversity that are most appropriately conserved in this manner.

Reserves cannot be the only strategy for maintaining our native biodiversity. Land used primarily for other purposes is likely to play an important role in protecting many genotypes, species and communities. If safety for biodiversity is positively related to population size and/or area occupied, any retention of native biodiversity on any land will improve the chances of avoiding extinction.

This chapter covers the problems associated with designing and managing protected areas in Australia by attempting to answer the question: 'How can we have a secure, viable and fully representative protected area system in Australia?' It briefly discusses the role of integrated land management in biodiversity conservation, and presents a synopsis of the current conservation status of Australia's major vegetation types and the problems involved in conserving them.

CRITERIA FOR PROTECTED AREA SELECTION

In 1974 Professor R.L. Specht and his associates argued for a national system of ecological reserves:

> A network of reserves, based on major ecosystems, would conserve a large proportion of the plant and animal species recorded in Australia. The network would focus longterm conservation objectives on those areas which would provide the greatest diversity of ecosystems, a wide range of plant and animal species, as well as a wide variety of heterogenous habitats in which evolution could proceed far into the future. In effect, this approach would channel limited manpower and resources into those conservation areas likely to achieve the greatest long-term benefit, rather than dissipate these energies on the emotional issues of conserving a single 'rare and endangered' species.

To provide a basis for the planning of these ecological reserves the plant communities of each state and territory were listed, to the extent allowed by imperfect knowledge, and their existing reservation assessed, again to the rather restricted limits of the information that existed in 1971. Plant communities, called alliances, were defined by height limits, the amount of shade cast by their tallest layer and by the most abundant species in their tallest layers. The selection of new ecological reserves would be directed towards those large areas containing the maximum number and variety of those alliances that were shown to be poorly reserved and unreserved.

The rationale behind concentrating on the reservation of alliances was that this was the only type of comprehensive information on the native flora and fauna that could be gained in a short time with a small number of workers, and that, if a variety of alliances were reserved, animal species and plant species would inevitably be reserved too. The logic behind this latter proposition is simply that the distributions of alliances, plant species and animal species are all ultimately controlled by the nature of the environment, so a reserve system that catered for any one of these biological elements could be expected to cater for the others.

In the late 1980s the above argument was taken a step further, as Professor Nix and his associates developed geographic information system methods to classify environments for nature conservation planning. These environmental domain analyses classify areas on the basis of climate, geology and topography. Notional reserves are then selected to cover the full range of environmental domains using proportional targets and rules related to contiguity.

Unfortunately, the few studies that have compared the predictive abilities of environmental domains and alliances have suggested that many of the rarer and more threatened species are not likely to be present in reserves planned on the basis of these coarse criteria. Environments cannot be good predictors of the ranges of those species that are absent from particular areas because of their migration history, or disturbance regimes that impinge upon the regeneration niche. They are good predictors of the more common and widespread species.

If the best means of planning locations for reservations is through the plant community, the alliance is a unit that should only be adopted where absolutely necessary for pragmatic reasons. The floristic composition of vegetation of the same height, projective foliage cover and species dominance can vary enormously. For example, there are very few species in common between the Scoparia (*Richea scoparia*) heath found near Federation Peak in the south-west wilderness of Tasmania, and the Scoparia heath found around poorly drained areas on the Central Plateau of Tasmania, just as there are few species in common between the Mulga (*Acacia aneura*) scrub in the north-western part of the Western Australian desert and the south-eastern part of the arid zone in New South Wales. Alternatively, vegetation that differs markedly in its structure and species dominance can be virtually identical in its species composition. For example, in many parts of Australia the combination of eucalypt species changes dramatically, while the species composition of the understorey responds less spectacularly to environmental gradients. In many vegetation types there are quite dramatic changes in structure and dominance that occur sequentially after fire or other disturbance, while species composition remains more or less constant. In tropical and subtropical rainforest the profusion and intermixture of tree species is such that the perception of alliances is extremely difficult.

If ecological reserves are to conserve the maximum number of native species, then the largest possible number of species should be included in the classifications that form the basis for choice of reserves. Thus, a classification based on lists of the bigger plant species would be better for species conservation than a classification based only on the dominants.

An expert observer does not take much longer to record a list of plant species from a site than to make a reasonable record of its structure, and the programs necessary to analyse such species presence and absence data are widely available on the computers owned by government and educational institutions. Such analyses do not initially have to tackle the monumental task of producing a classification based on species for all Australian vegetation. Studies of major vegetation types can ultimately be built into comprehensive classifications for states, regions and the Commonwealth, but even studies of the floristic variation of vegetation types within regions can provide better priorities for nature conservation than would otherwise be available. Of course, they can only do this if the national deficiencies in reservation are known, through studies such as that on the reservation status of alliances.

Despite some criticism directed towards the concentration of resources on preserving rare and threatened species, these must be incorporated into reserve planning if biodiversity is to be maintained. At the simplest level, it is sensible to reserve an area of an ecological community with rare and threatened species rather than an almost identical area without them.

Our present reserve systems might best be expanded or modified with the aim of ensuring the long-term reservation in the wild of as many genotypes of as many native species as can be and need to be perpetuated in this manner, rather than with the direct aim of making large reserves that contain as many domains, alliances or floristic communities as is possible. Given enough time, and appropriate environmental conditions, alliances and communities will reconstitute themselves, but only as long as the species that dominate them survive. If we concentrate on reserving the alliances that exist today and abandon rare species to their fate, we may be encouraging the destruction of the dominants of the alliances of tomorrow, when conditions change to make rare species abundant. Species that are isolated genetically form the most urgent targets for reservation and other appropriate conservation measures.

Concentrating on the perpetuation of species also has the virtue that it allows the use of cultivation and transplantation; methods that cannot be used to preserve alliances in any practical way. There is no doubt that most species are real, in the same sense that most individuals of most species are distinct and real entities, whereas alliances are but convenient abstractions. The reality of species makes them the appropriate units on which to base reservation strategy.

The present reserve system contains large populations of many variants of most Australian native plant species. Only one hundred or so

Australian native higher plant species are thought to have become extinct since European settlement. Another two hundred odd species are considered to be in sufficient danger of extinction to be labelled endangered. If the minimum satisfactory reservation for a plant species were considered to be two large and well-separated populations, approximately 400 reserves would be required for the two hundred species.

However, rare or threatened species tend to aggregate, often within very small areas. Thus, the areas of highest priority for reservation should be those with the greatest numbers of considerable populations of the greatest number of unreserved plant species.

SIZE OF RESERVES

The inclusion of a variety of plant communities within large National Parks is an admirable aim in itself, and is certainly beneficial for nature conservation and the growing band of wilderness recreationalists. However, there are strong arguments against these large reserves being the principal focus of all serious long-term nature conservation activity.

Minimum population size

The advocacy of large reserves rests on four major arguments. The first relates to the minimum population size necessary to ensure the survival of a species and its gene pool (the minimum viable population). As environments tend to fluctuate quite markedly, a normal population necessary to ensure the survival of a species and most of its gene pool will be many times greater than the minimum population below which genetic attrition or extinction would inevitably result. A study of the genetics of lizards on Californian offshore islands that had been isolated from each other for 15,000 years suggested that the avoidance of genetic depletion for lizards in that environment over that time period required normal populations in the order of 10,000 individuals. Geneticists who have pondered the problem, tentatively suggest that populations of fifty or more are necessary to maintain genetic fitness, and that populations of 500 or more are necessary for the longer term maintenance of variability. Whatever the exact levels, and these will certainly vary by organism and circumstance, it is maintained that species with small populations, all else being equal, are more likely to become extinct than those with large populations. Therefore, the minimum size of a reserve should be controlled by the area needed to support a viable long-term population of the species with the greatest requirement of land per individual.

The above strategy makes for large reserves, especially in the non-productive country in which they are usually located. For example, the

Red Kangaroo has become extinct on Barrow Island (200 km^2) in the ten thousand or so years since it parted company with the mainland desert, while a population of 2000 euros has survived. If Red Kangaroos had survived there would be 200–300 of them on the island. If we assume that a normal population of 2000 Red Kangaroos is necessary for the long-term survival of the species, a reserve of at least 2000 square kilometres would be required. Of course, like many other animals whose food requirements are not species specific, the Red Kangaroo happily utilises land totally transformed by human activity.

A more recent approach to determining the likelihood of extinction than the minimum viable population approach, calculates probabilities of extinction based on the biology of the organism and assumptions about environmental impacts on metapopulations (collections of related populations). This population viability analysis provides extinction probabilities at different timescales for different current population numbers and spatial arrangements. It is an expensive process that must ultimately rely on many assumptions, so has only been used for organisms with a high public profile or of key ecological importance. The practicalities of reserve planning do not require such analysis for most species. If reserves contain large populations (10,000 plus individuals) of a species within spatially separated populations, all previous work suggests that the species is as safe as it can be made through the simple act of reservation. Analysis is most needed for those species with low total numbers or low numbers in reserves.

Large ecological reserves are not necessary for maintaining minimum populations of most space-requiring animals, as long as they are not managed to extinction on land primarily devoted to other purposes. What is more likely to be needed is the reservation and management of small areas that are critical to the survival of a species, and the promulgation of regulations that control human predation wherever necessary to ensure the maintenance of desirable population levels. For example, the Orange-bellied Parrot has been reduced to a total population of less than two hundred birds, whose breeding grounds within the South West National Park, an archetypal ecological reserve, are more than well-provided for. The bird migrates to Victoria and South Australia via the west coast of Tasmania and King Island where it feeds off saltmarsh plants and the introduced Sea Rocket, coloniser of beach strands. It needs small well-managed saltmarsh reserves on the mainland, a small reserve on King Island and absolute protection from human predation. Most plants are distinctly less mobile than most animals, and therefore require much smaller reservations to ensure safe population levels, although their inability to flee catastrophes makes replication of reservations even more important than it is for animals.

The theory of island biogeography

The second argument for large reserves is closely related to the first. It is based on the theory of island biogeography. Many reserves are, or may become, islands of native vegetation in a sea of development. The theory of island biogeography would seem to suggest that there will be an inevitable loss of native species from these islands habitats, and that the size of this loss will increase as reserves become smaller and as they become further isolated from other such habitat islands. The large numbers of seeds of introduced plants that would penetrate a habitat island reserve from the surrounding cleared land would be likely to swamp the much lesser amount of seed that could be expected to reach the reserve from other habitat islands of native vegetation. Thus, the equilibrium number of native species is likely to be severely depressed below that which could be expected for a real island, as the sea does not act as a source of disseminules of land plants.

Some scientists have maintained that the theory of island biogeography supports the proposition that a large island is likely to ensure the long-term survival of more species than a number of smaller islands that add up to the same area. This proposition is only true theoretically when highly specific characteristics are assumed for the species in the pool available for immigration and extinction. In reality the proposition is almost never correct. The natural distributions of most species are not of uniform density, and environment, vegetation and species composition seldom remains constant over large areas. Thus, a species that is exceedingly rare in one climatic extreme of the distribution of an alliance may be exceedingly common at the other extreme, and vice-versa. A series of small reserves could be designed to include substantial populations of all species along an environmental gradient, whereas a single large reserve of equivalent area and compactness could not hope to do so. Figure 8.1 illustrates this reality in a simple case. Species distributions are shown as overlapping, bell-shaped curves, which imply, as is supported by much ecological research, that the species composition of vegetation changes gradually as the environmental conditions change, and that each species has a position on the environmental gradient at which it is most abundant and from which its numbers decline in both directions. Any non-contiguous configuration of small reserves along this gradient will include more species than a single large reserve of equivalent size, even if species of low abundance within a reserve are assumed to be doomed to local extinction. In the particular case shown in Figure 8.1, the six small reserves contain

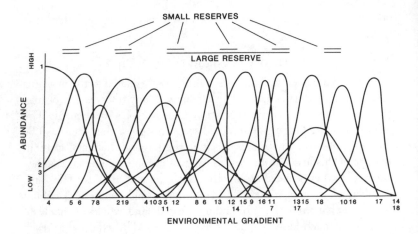

Figure 8.1 The effects of several small versus a single large reserve on species conservation over an environmental gradient.

all eighteen species at the population levels assumed to be sufficient to ensure their long-term survival, whereas the areally equivalent large reserve contains only eleven species.

Evolutionary continuity

The third argument for large reserves is that their size and heterogeneity will be necessary to ensure the evolutionary continuity of our native biota. It is basically a plea to preserve as much as possible of the variable genetic inheritance of species, in order that the raw material for further adaptation and evolution should be available for future environmental vicissitudes. The plea is irreproachable, but are large reserves necessarily the way to satisfy it?

A more convincing argument for large reserves for evolutionary continuity is that they allow environmental space for the retreat or expansion of species in response to inevitable environmental change (Chapter 3). The magnitude of past changes in climate and vegetation gives us some indication of the scale of change that might occur in the future. Both temperatures and precipitation plummetted over large areas of Australia during the Last Glacial. In parts of Australia these changes were so great and so extensive that species could only survive by massive shifts in their distribution patterns, shifts that are verified by the pollen record and that may, in some cases, still be taking place. Many species would have been much more widely distributed than today during the cold of glacial times. During the present warmth their

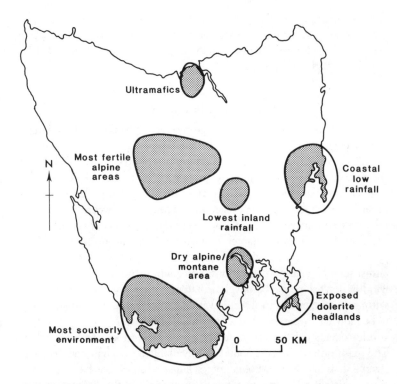

Figure 8.2 Centres of local higher plant endemism in Tasmania (generalised and updated from Kirkpatrick, J.B. and Brown, M.J. (1984)). The palaeogeographic significance of local endemism in Tasmanian higher plants. *Search* 15, 112–113).

distributions have contracted to refugia, each of which characteristically contains many such species. For example, Figure 8.2 shows the areas in which two or more local Tasmanian endemic species are found. The greatest concentrations of these restricted endemics occur where conditions are most similar to those that are thought to have been most widespread during the Last Glacial. Many of the species that are presently widespread in Australia must have had similarly restricted distributions during cold and/or dry periods. Unless our reserve system and land use patterns allow the species to move along their previous lines of retreat they will be doomed to extinction. Thus, for many species and communities, their evolutionary survival might relate more to the degree to which the locations, configurations and environmental heterogeneity of reserves allow retreat than to reserve size. In most cases large reserves could be necessary to ensure retreat routes. In other cases small stepping stone reserves might be all that is possible, but these reserves could have greater long-term significance than an alternative compact large reserve in an area of uniform environment and little palaeogeographic significance.

The problem of edge effects

The fourth major argument for large reserves is based on the changes that occur, and the restrictions on management that apply, near the boundary of reserved bush with city, farm or industrial forest. The larger the reserve, the smaller will be the proportion of its area subject to these edge effects. The degree to which edge effects necessitate large reserves differs considerably between vegetation types and between locations in relation to other types of land use. The potential effects of nutrient drift and fungal attack (see Chapter 7) make the vegetation types found on poor soils ill-suited to small reserves that have common boundaries with farms, suburbs or pine plantations, especially if the reserves are downwind or downslope of the developed areas. Other vegetation types, like saltmarsh, mangroves, freshwater wetland and savannah woodland seem to be more easily perpetuated within small areas.

Edge effects can be minimised by good reserve location and design. The sea is one of the best boundaries available for a small conservation reserve. It is also a good boundary for larger reserves, but has the disadvantage that it allows ready access to an element of the population that indulges in rubbish dumping and fire lighting, unless the land/sea interface consists of rugged cliffs. Steep cliffs and other rocky areas generally make good boundaries. However, away from the coast the most important attributes for small reserves containing ecosystems susceptible to edge effects are that they contain the headwaters of their catchments, and are abutted by land used for such extensive purposes as rough grazing or logging. Wherever possible they should be in positions that neither risk the spread of fire from other land, or menace other land with fire. All else being equal, the more nearly reserve boundaries approximate a circle, the less will be the percentage of the reserve subject to edge effects. However, circles can only be recommended for featureless plains, most boundaries being best situated on ridges, cliffs, lakes or big rivers to minimise edge effects.

The above discussion and the reality of the fragmentation of many native ecosystems suggest that ecological and small reserves should be incorporated in a system of reserves planned to maximally represent biodiversity.

LOCATION OF RESERVES

National Parks and other protected areas are a twentieth century land use in Australia. Until the second half of the century, almost all reserves were chosen because of their scenic grandeur, their recreational potential and their economic uselessness. In the last few decades of the century, wilderness recreation has impelled much of the reserve network expansion, and representativeness of environment and biodiversity have also become prominent. The early techniques for choosing new reserves on biological grounds scored species and communities according to such attributes as their rarity and reservation status, added up the scores for the areas studied and determined priorities on that basis.

In the early 1980s I realized that this process was likely to result in areas with the same species and community complements being given highest priority for reservation, and, consequently, a reservation system that left out those significant taxa that did not co-occur with other significant taxa. In a study of Tasmanian endemic plant species and their conservation needs on the central east coast of the state I developed an iterative method for selecting nature reserves.

Each square kilometre of the study area was given a score that gave greater weight to the unreserved than the poorly reserved species. The grid square with the highest value was selected for the core of the first priority reservation. It was then assumed that this reserve actually existed and all the grid square scores were recalculated on that basis. The grid square with the highest score was then selected as the core of the second priority reserve and the process repeated. Only six reserves were necessary to cover almost all of the species.

Since the early 1980s, minimum set techniques, which do not indicate priority, have been developed by a group of scientists in New South Wales. These techniques set out minimum reservation criteria and select the minimum area of land of defined tenure and contiguity that will satisfy them. Recent refinements of their technique include the capability of determining the replaceability of any selection.

We now have reserve selection techniques that enable optimal and space-efficient location. The value of these techniques depends upon appropriate assumptions for minimum reservation and adequate biological distribution data. Our data on the distribution of plant species and communities may never be complete, but the more complete it is the more credence we can give to the conservation value of a reserve system selected by these means.

People Management Problems

Large reserves present fewer management problems than small reserves. The present allocation of resources to the parts of the public service entrusted with the management of National Parks and other conservation reserves is so meagre that only the largest and most popular of parks enjoy the constant presence of rangers. Large reserves are able to suffer the impacts of recreational use without totally destroying their utility for nature conservation, whereas many smaller reserves are highly vulnerable to recreational destruction, especially in the absence of either policing or a positive community attitude.

People present the major management problem for conservation reserves, as well as being their ultimate reason for existence. In some cases there are dangers that reserves will be loved to death. In other cases they are damaged by ignorance, greed or malice. Many rural people tend to regard the bush pragmatically. They see it as a source of weeds and vermin, a fire danger, a hunting ground, a source of wood, a place to rough graze stock and a place to dump rubbish. Some of them will continue to use nearby bush in their accustomed way until absolutely forced to desist. They will certainly continue to agitate for control of the dangers to their own interests that they perceive to emanate from a reserve.

At the insistence of the local graziers, a large reserve in western New South Wales was fenced to keep kangaroos from private pastures. This fencing was fortunate because the pastures are now infinitely superior on the reserve side of the fence. Hazard reduction burning of bush near reserve boundaries with private land has been a less fortunate form of compliance to local pressure.

As if the local inhabitants did not present enough problems for conservation reserve management, the last decade has seen a proliferation of urban-based mechanised adventure recreationalists. The high pitched screams of the trail bike and motorboat and the growls and screeches of the four-wheel drives and dune buggies are now as much a part of the Australian bush as the laugh of the kookaburra and the dry rustle of the gum tree leaves. Off-road vehicles tend to be of the most use in the very environments that are least likely to be able to cope with their impact. Where growth rates are fast and the trees and shrubs are dense they can gain access only along already formed tracks, but in the extreme environments of salt marsh, sand dune, heath, alps and desert they find little impediment to their progress.

Off-road vehicles cause breakage, abrasion and compaction. The deaths of plants and parts of plants that result from breakage and abrasion can be compensated for by growth if the traffic is infrequent enough and the destruction of vegetation has not resulted in accelerated erosion. The soils of the coastal sand dunes, deserts and alps are particularly prone to movement by wind and water, once exposed to the weather. The fact that the weight of vehicles compresses the soil also contributes to the likelihood of water erosion. Compression reduces the space between the soil particles, thereby reducing soil's ability to hold water, and reducing the possible infiltration rate of water into the soil. These two changes mean that more water is likely to run off the surface of the soil than was the case before compression. Therefore, it is possible for more erosion to occur. This tendency is exacerbated by the channels that compression creates and by the preference exhibited by most off-road vehicle enthusiasts for the most direct routes up and down hill. After heavy rain the off-road vehicle is more than likely to create a string of wallows as it negotiates its way across the countryside. These wallows tend to expand as each driver attempts to avoid quagmires by driving over unbroken vegetation. Similar features can be created in dry, sandy country once one vehicle breaks through the surface layer of roots.

The off-road vehicle enthusiasts like the bush because they are free of the restrictions of the road and can become pioneers of new routes in an adventurous and challenging situation. Although some people who use such vehicles are models of environmental rectitude, most have a disproportionate share of the attitudes to moving and stationary organisms in the bush that made the gun, the axe and the match the great Australian tools. Their camping places and huts can be recognised by the piles of tubular tin and brown glass artefacts and the axed and charred remnants of bushes and trees. Their attainment of remote goals is usually celebrated in paint or sculpture upon some prominent feature. They are much more likely to start bushfires than the tame tourists encapsulated in campervan and kombi on the main road, and they are usually found in the places where such fires can do most harm. Their independence of spirit makes the locked gate a challenge to be met with the winch or a rifle shot.

Because there are many legitimate uses of off-road vehicles it is difficult to curb their sale. It has even proved difficult to enforce their compulsory registration, making great difficulties for those who wish to apprehend the more offensive users. Some areas have been abandoned to off-road vehicle activities in the hope of concentrating damage. Sand dunes, bared by other causes, are a promising locality for these mechanised recreation parks, as the vehicles can do little to exacerbate erosion problems if there is no vegetation to destroy.

The use of recreational off-road vehicles is forbidden in, if not absent from, most Australian conservation reserves. However, fishing is allowed in most reserves, usually after the legal introduction of a highly competitive introduced species, the trout. Many National Parks are also the sites of holiday villages, some of which are devoted to the sport of downhill skiing. These villages demand the usual urban amenities of sewerage, power, access and rubbish disposal, all of which involve the destruction and alteration of bush. The use of bulldozers, chainsaws and explosives to create suitable ski runs inevitably alters more bush. Even the weight of people on skis is capable of compacting the soil beneath the snow, although not to the same extent per person as the boots of walkers. Where intensive recreational development takes place within National Parks there is a tendency for the park managers to allow it to become the focus of their activities. The managers and owners of the considerable businesses based on National Park villages become a powerful political lobby, and other businessmen wish to get a piece of the action, leading to constant expansionary pressures.

Outside the perceived snow zone, tourist accommodation has been increasingly supplied on the margins of National Parks. Some old tourist villages, such as that near Uluru, have been dismantled in favour of less obtrusive and destructive facilities, and caravan parks are now usually placed on the margins of parks or well outside them.

The tourist villages and caravan parks are the most intensively used parts of the conservation reserves in which they occur. They often form the node of a network of walking tracks. The shortest tracks are usually constructed in a manner that allows access by the very old, the very young and the infirm. They are also usually characterised by numbers on posts related to descriptions in pamphlets, or by tasteful metal signs giving equivalent information, and by unobtrusive wooden constructions that guide to the best vantage points and discourage exploration off the track. Tracks attract fewer people the further they extend from a car park, and the less they are aimed at a spectacular goal. Distance from carparks also increases the cost of construction and maintenance, so unobtrusive dark wooden constructions and smooth, easily-walked surfaces are seldom found beyond the nature trail. Most walking tracks more than a kilometre from a car have their present location determined by the path of least resistance taken by the first walkers to explore the area.

The path of least resistance is usually the worst possible location for a walking track used by more than a few people per year. The exploratory walker will prefer treeless or thinly treed areas to dense forest and will climb hills and mountains by the most direct route perceivable. A lack of trees often denotes poor drainage, and poorly drained tracks quickly develop into rivers of mud with even moderate use. Most people who have walked the overland track through the Cradle Mountain – Lake St. Clair National Park in Tasmania have had the experience of disappearing up to their crutch in mud at least once. Such unpleasant experiences suggest to most walkers that they might be more comfortable trampling the intact vegetation on the sides of tracks than risking their all in the quagmire. The resultant morasses only cease to widen when walkers gain well-drained ground, or when the centre line of the morass has been abandoned so long that plants are able to colonise, and thereby attract back walkers. The path directly up a slope is a recipe for accelerated erosion. Most become intermittent streams flowing in gullies of varying depth. Such gullies form even on gentle slopes as long as they effectively channel water. The only tracks straight up slopes that do not cause such erosion are those on rock. The number of walkers per year whose passage will create widening wallows or deep gullies, rather than evident but unobtrusive pads, varies according to environment. Some environments would be totally impervious to the daily passage of armies. Beaches, dry riverbeds and anywhere that rocks form a substantial portion of the land surface fall within this class. Anywhere else, the number of feet needed to cause accelerating damage depends on the growth rates and adaptations of the most trampling-resistant species in an area.

The adaptations that minimise trampling damage include elasticity, the ability to grow from below rather than from the tips of foliage, and the ability to adopt a prostrate habit. Rapid growth rates may allow well adapted species to persist on quite heavily used tracks, like the grass on cricket pitches. Thus, vegetation destruction through trampling is more likely to be seen in extreme environments, where growth rates are depressed through lack of nutrients, lack of warmth or a lack, or excess, of water.

Some indication of the threshold levels for accelerating damage in different environments can be gained from studies relating track environment to track usage. A study in the Cradle Mountain – Lake St. Clair National Park found threshold levels to be as low as 500 people per year in poorly-drained, oligotrophic, alpine environments. Yet over 2000 people per year used the Overland Track, which passed through much of this sort of country. The inexpensive solutions to the vegetation destruction problem caused by trampling are exclusion, rationing and relocation. Access to National Parks has already been rationed where the number of people wishing to use camp sites has exceeded their capacity. For example, it is no longer legally possible to set off on a camping trip in

Plate 8.1 Duck boards constructed on a highly degraded track.

Wilsons Promontory National Park without going through an application procedure that might well end in refusal. Walkers have been excluded from some conservation reserves and from some parts of others in order to avoid problems with nature conservation.

Both exclusion and rationing will probably have to be used more in the future, as the increasing number of people who wish to recreate in the bush exceed the capacity of some popular parts of the bush to cope with their impact. However, in many situations track relocation can prevent or rectify recreational damage. The opportunities for relocation are greatest where the old track lies adjacent to better-drained and more rocky country. However, with tracks of any great length the total avoidance of vulnerable ground is likely to be impossible. Frequent relocation of track alignment on vulnerable ground is not usually a satisfactory solution, given the low numbers of feet that will create perceptible damage, and the not much greater numbers of feet that will destroy all vegetation. Relocation also does not help much where the process of accelerated erosion has already been induced on the old track, because such erosion, once started, will continue, without another foot being placed on a track. Thus, more expensive solutions become necessary.

Where tracks are potential quagmires the walker can be separated from the ground by duck boards (Plate 8.1) or metal mesh, the track can be gravelled down to firm ground, or pieces of wood can be used to cord the track. Duck boards, metal mesh and gravelling can be as expensive as building a road. Duck boards are also easily broken by a heavy weight of snow, making them unsuitable for the alpine zone on the Australian mainland, and the treated pine does not last for much more than a decade, requiring frequent maintenance and reconstruction. Both gravel paths and duck boarding tend to be much more conspicuous than metal mesh or cording, although not nearly as conspicuous as unimproved and widening tracks bared down to bedrock. Cording requires only a source of wood and the labour of a few men, as opposed to the expensive helicopter airlifts that make other forms of track construction so expensive. Its cheapness is reflected in its comfort for the walker, which is minimal compared to the more expensive methods, but better than wallowing.

Eroding tracks allow no cheap alternatives. The key to their long-term restoration is to divert the erosion agent. The erosive power of water is proportional to the cube of its velocity, the velocity being a function of gravitational force and the degree of restriction of a stream. Erosion is also related to the amount and duration of flow. Amounts, durations and speeds can all be reduced by cross drainage works that break up a steep track into small and gently sloped catchments which debouche broadly into the adjacent untrampled vegetation. If this work is combined with the construction of steps by cutting and/or infilling, a moderate amount of maintenance can ensure that further severe erosion is prevented.

As well as having an impact on tracks, bushwalkers can also cause considerable damage in the areas in which they camp. Apart from nutrient accessions from urine, camping areas tend to accumulate rubbish, be surrounded by ground dug for defaecation, be encircled by a zone almost totally lacking in dead wood, and to be centred on black circles tastefully garnished with silver paper and backed with monumental rock constructions. Many unplanned bush fires are caused by bushwalkers who do not believe, or care, that peat can burn or that untended fires can make their way to nearby dry plant litter.

Only the impacts impelled by normal bodily functions are unavoidable, and these impacts may encourage a greater dispersal of camp sites than exists at the present. The more crowded camp sites on the rivers and tracks of the remaining wilderness areas in Australia are becoming sources of gastro-intestinal diseases. The use of natural fuel for cooking ends up in the destruction of living plants and creates an unnecessary fire hazard. Well-frequented camp sites and warm, dry weather demand the use of portable stoves, powered by fossil fuels or alcohol. Anything that becomes rubbish after use should be carried out. There is no need for monuments to human endeavour to arise in the bush.

Despite many localised disasters the effects of bushwalkers, or even off-road vehicle enthusiasts, are not, at the moment, seriously threatening the existence of any native species and ecosystems. However, if the bush continues to retract while bush recreation grows, the outlook will not be so sanguine.

A SYNOPSIS OF THE STATE OF VARIOUS VEGETATION TYPES

Rainforest

Between one half and two thirds of Australia's rainforests have avoided clearing. A high percentage of the remaining area is in conservation reserves, although gaps exist with the communities of drier country and there are many species endangered by past clearing. Fire is a major threat to the survival of temperate rainforest, which is more easily burned than most other types. Introduced vines, particularly *Thunbergia grandiflora* and *Cryptostegia grandiflora*, form a threat to tropical and subtropical rainforest.

Alpine vegetation

Very little alpine vegetation has been lost to clearing and inundation. Most of the area of alpine vegetation is in conservation reserves, although, in Victoria, this does not preclude stock grazing, which has been shown to change the vegetation and reduce the populations of some native species. Fire has degraded a large proportion of the alpine vegetation of Tasmania, leading to local extinctions of species and communities. Its exclusion from alpine vegetation presents a major management problem. Exotic species are rare in Tasmania and integrated into the vegetation on the mainland. There are no major threats from this source. Damage from recreational activity is widespread. There need to be stronger limits on skiing, bushwalking and horse-riding and a total exclusion of off-road vehicles. Substantial climatic warming would threaten many alpine communities and species.

Wet eucalypt forest (wet sclerophyll forest/tall open-forest)

Much of the area of this vegetation type was cleared for agriculture. The reservation status of the type is moderate, although many communities are poorly reserved. Fire management problems have been discussed earlier. Several shrubs and trees may invade this vegetation type in the

Plate 8.2 Alpine vegetation at Mt Bobs, Tasmania (A. Bowden).

long term. However, at present there is no exotic species that is threatening the future of any wet forest species or community. Much of the remaining wet forest is dedicated to logging (see Chapter 5 for a discussion of the implications).

Dry eucalypt forest (dry sclerophyll forest/open-forest)

This heterogeneous vegetation type has suffered, and is suffering, massively from land clearance. It contains many threatened plant species. The reservation status of the types on poor soils is generally good, but the reservation status of types found on better soils is usually poor. The Cinnamon Fungus threatens the species richness of large areas. Brooms, Boneseed and Gorse are some of the many invaders that threaten large parts of the dry forest estate. Introduced grazing animals and stock present a major management problem. These forests are widely used for wood production (Chapter 5).

Tropical eucalypt forest

Very little of this vegetation type has been cleared, although there have been constant attempts to establish improved pastures at the expense of the trees. The reservation status of tropical eucalypt forest is moderate to good. The major threat to the future of this vegetation type is the introduction of new pasture species and the invasion of Mission Grass (*Pennisetum polystachion*).

Temperate grassy woodland and grassland

The greater part of the area of this vegetation type has been cleared or severely degraded. Temperate lowland grasslands have been reduced to much less than 1% of their original area. The decline continues apace. A large number of threatened plant species are found in this type. Reservation status is extremely poor. Gorse and *Stipa neesiana* are the two worst of the many introduced species that threaten to transform the pitiful remnants from clearing. The remnants need protection from overgrazing, but usually require biomass reduction through grazing and/or burning to maintain the intertussock species.

Mallee

A large proportion of the original area of mallee has been cleared for wool and wheat production. This clearing continues in New South Wales and Western Australia. The reservation status of mallee is moderate to good. Introduced animals, particularly the rabbit, may threaten the future of the mallee. Inappropriate fire regimes can lead to the elimination of the eucalypts. Some mallee communities on silicious soils in areas with moderate rainfall are likely to lose part of their species complement as the result of the invasion of Cinnamon Fungus.

Mitchell (*Astrebla*) grasslands

These have been all subject to stock grazing. Until very recently, there

Plate 8.3 *Eucalyptus pauciflora* grassy woodland and grassland (S. Harris).

have been no reserves and no plans for reserves. The ecosystem is threatened by the invasion of an African shrub, *Acacia nilotica*.

Heath

Large areas of heath have been cleared, particularly in the Western Australian wheat belt and South Australia. Clearance continues in most states, but at a lesser rate than in the past. The remnants in Western Australia have an extraordinarily high concentration of threatened plant species. The major threat to the future of this vegetation type is the invasion of Cinnamon Fungus, which threatens the extinction of a wide variety of species (see Chapter 7). Nutrient enrichment leads to the rapid demise of heath.

Desert vegetation

The hummock grasslands, chenopod shrublands, *Acacia* shrublands and herbfields of the Australian desert have been deleteriously affected by stock and feral animal grazing and changes in fire regimes, especially in the temperate zone. There are many large reserves in the arid zone, but few that extend over major precipitation gradients. The control of grazing pressure is critical to the future luxuriance of our desert vegetation. The vegetation in and around sources of fresh water is the highest priority for protection. The introduced grass, *Cenchrus ciliaris*, and the introduced shrub, *Tamarix aphylla*, threaten the ecological integrity of large portions of the arid zone.

Wetlands

A large proportion of the wetlands of temperate humid Australia has been drained, inundated or starved for water. The tropical and arid zone wetlands are almost intact, although the mound springs in the arid zone, which rely on artesian water, have a precarious future. Reservation status is highly variable, depending on the type of wetland. Wetlands are threatened by a wide variety of introduced plants. The most notable are: *Mimosa pigra*, a prickly shrub that invades tropical flood plains; *Eichhornia crassipes*, a floating herb that fills open water; *Brachiara mutica*, *Echinochloa polystachia* and *Hymenachne amplexicaulis* – all tropical grasses introduced for wetland pastures; *Glyceria maxima*, a temperate wet pasture grass; and *Salvinia monesta*, an occupier of open water capable of changing hydrological systems.

Coastal vegetation

Much coastal vegetation has been destroyed by development or reclamation. Reservation status is moderate to good depending on the type.

Coastal sand dune and cliff vegetation is highly susceptible to invasion by introduced plants. Boneseed and Marram Grass (*Ammophila arenaria*) are two of many species that have transformed coastal vegetation after escaping from rehabilitation plantings. Saltmarsh has proved resistant to exotic invasion, although the introduced grass, *Spartina townsendii*, has occupied its lower reaches in some parts of south-eastern Australia. The combination of firing and grazing leads to a loss of some elements of the saltmarsh flora. Mangroves do not appear to be subject to any invaders, and are subject to no subtle threats apart from the possibility of being squeezed between development and the sea as a result of sea level rise.

INTEGRATED LAND MANAGEMENT

It seems unlikely that all Australian natural and semi-natural vegetation will be placed in conservation areas. The large proportion of the continent used for extensive stock-grazing, logging and other economic activities based on native vegetation can be managed to maximise biodiversity conservation within the constraints of enterprise.

The idea of bioregions, recently promulgated through a variety of policy documents, including those emanating from the ESD process, is to make the planning process congruent with ecological conditions. Thus, bioregions are characterised by their relative biological uniformity, as opposed to existing political and planning divisions that have boundaries relating to historical decisions and economic networks. Integrated catchment management recognises the dependency of the conditions downstream on those upstream. These approaches could benefit biodiversity conservation by allowing planning to encompass all land, not just that in the protected area system. For example, such planning could mitigate edge effects and maintain connectivity between reserves for at least some species.

Our vast area of semi-natural grazing land in the arid zone, the semi-arid zone and the wet/dry tropics should be subject to sustainable ecologic management which maintains productivity for stock based on the perpetuation of native species. Such management is made difficult by the highly unpredictable climate and the optimistic attitude of most graziers. A good scientific basis for sustainable ecological management exists. It remains only to ensure that the economic incentives and legal limitations provide a motive for its implementation.

References and further reading

(Most of the references given below have been used by the author in writing this book.)

Chapter 1

The subject of biodiversity is outlined by the Biological Diversity Advisory Committee (1992), *A National Strategy for the Conservation of Australia's Biological Diversity — Draft for Public Comment*, AGPS, Canberra. More detail is available in Amos, N., Kirkpatrick, J.B. and Giese, M. (1993), *Conservation of Biodiversity and Ecological Integrity and Ecologically Sustainable Development*, Australian Conservation Foundation, Melbourne.

Extinction and endangerment in Australia are reviewed in Kirkpatrick, J.B. (1991a), The geography and politics of endangerment in Australia, *Australian Geographical Studies* 29, 246–54.

Ecologically sustainable development is well covered by Hare, W.L., Marlow, J.P., Rae, M.L., Gray, F., Humphries, R. and Ledgar, R. (1990), *Ecologically Sustainable Development*, Australian Conservation Foundation, Melbourne.

Eckersley, R. (1991), The concept of sustainable development. In Behrens, J.M. and Tsanemyi, B.M. (eds) *Our Common Future*, Faculty of Law, University of Tasmania, Hobart, pp. 46–57 gives an excellent critique of the concept of sustainable development, which is promoted by the World Commission on Environment and Development (1990), *Our Common Future*, Oxford University Press, Melbourne.

The following references are useful in considering the rights of the rest of the living world: Fox, W. (1990), *Towards a Transpersonal Ecology: Developing New Foundations for Environmentalism*, Shambhala Publications, Boston and London; Gunn, A. (1980), Why should we care about rare species?, *Environmental Ethics* 2, 17–37; Naess, A. and Rothenburg, D. (1989), *Ecology, Community and Lifestyle*, Cambridge University Press, UK.

Hypothesis testing in ecology is discussed in Quinn, J.F. and Dunham, A.E. (1983), On hypothesis testing in ecology and evolution, *American Naturalist* 122, 602–617.

Chapter 2

The best available general text in the area of this chapter is Kellman, M.C. (1980), *Plant Geography*, 2nd edn, Methuen, London.

Hnatiuk, R.J. (1990), *Census of Australian Vascular Plants*, Australian Flora and Fauna Series No. 11, AGPS, Canberra, provides a useful list of current scientific names for higher plants in Australia.

The role of climate in influencing plant distributions is described in detail in Stoutjesdijk, Ph. and Barkman, J.J. (1992), *Microclimate, Vegetation and Fauna*, Opulus, Sweden.

Attiwill, P.M. and Leeper, G.W. (1987), *Forest Soils and Nutrient Cycles*, Melbourne University Press, Melbourne, is the best source for information on the ecological impact of soils in general.

The classical book on the ecology of fire is Gill, A.M., Groves, R.H. and Noble, I.R. (1981), *Fire and the Australian Biota*, Australian Academy of Science, Canberra. Jackson, W.D. (1968), Fire, air, water and earth – an elemental ecology of Tasmania, *Proceedings of the Ecological Society of Australia* 3, 9–16, remains the most useful single paper for understanding the interactions between fire and the rest of the ecosystem.

Howe, H.F. and Westley, L.C. (1988), *Ecological Relationships of Plants and Animals*, Oxford University Press, New York, covers the general subject of biotic influences on plant distributions in considerable detail. The ruderal/tolerator discussion follows Grime, J.P. (1979), *Plant Strategies and Vegetation Processes*, Wiley, New York, while the cliff example is from Coates, F. and Kirkpatrick, J.B., 1992, Environmental relationships and ecological responses of some higher plant species on rock cliffs in northern Tasmania, *Australian Journal of Ecology*, 17, 441–449.

Morrow, P.A. (1977), The significance of phytophagous insects in the *Eucalyptus* forests of Australia, in W.J. Mattson (ed.), *The Role of Arthropods in Forest Ecosytems*, Springer, New York, pp. 19–29, reviewed damage to eucalypt leaves from predators.

The following references are useful on the regeneration niche and vegetation dynamics: Connell, J.H. and Slatyer, R.O. (1977), Mechanisms of succession in natural communities and their role in community stability and organisation, *American Naturalist* 111; Grubb, P.J. (1977), The maintenance of species richness in plant communities: the importance of the regeneration niche, *Biological Review* 52, 107–45; Noble, I.R. and Slatyer, R.O. (1980), The use of vital attributes to predict successional changes in plant communities subject to recurrent disturbance, *Vegetatio* 43, 5–21; White, P.S. (1979), Pattern, process and natural disturbance in vegetation, *Botanical Review* 45, 229–299.

Classical succession theory is outlined in Kellman, M.C. (1980), *op. cit.*, and most other plant geography text books. There is a good discussion of the role of dispersal in influencing plant distributions in Smith, J.M.B. (1982), *A History of Australasian Vegetation*, McGraw-Hill, Sydney. MacArthur, R.H. and Wilson, E.O. (1967), *The Theory of Island Biogeography*, Princeton University Press, Princeton, started it all.

Sabath, M.D. and Quinnell, S. (1981), *Ecosystems — Energy and Materials*, Longman-Cheshire, Melbourne, is an Australian-oriented text on the subject of ecosystems.

Gleick, J. (1987), *Chaos*, Sphere, London, is an accessible treatment of this subject. Gillison, A.N. and Anderson, D.J. (1981), *Vegetation Classification in Australia.*, CSIRO and ANU Press, Canberra, covers the problems and procedures of categorization. Kirkpatrick, J.B. and Dickinson, K.J.M., (1986), Achievements, conflicts and concepts in Australian small-scale vegetation mapping, *Australian Geographical Studies*, 24, 224–34, discusses the interface between classification and mapping and the nature of boundary conditions. The classification of Specht can be found in Groves, R.H. (1981), *Australian Vegetation*, Cambridge University Press, Melbourne.

Grime (1979) *op. cit.* and Connell, J.H. (1978), Diversity in tropical rainforests and coral reefs, *Science* 199, 1302–30, provide useful discussions of the causes of variation in species richness.

Chapter 3

Dodson, J. (1992), *The Naive Lands*, Longman Cheshire, Melbourne, provides the most recent and most comprehensive account of the relationships between Aborigines and the vegetated landscape. The work of Singh on Lake George is fully described therein and also in Gill *et al.* (1981) *op. cit.*. The recent work of Jones and Kershaw and their colleagues referred to in the text is described in a series of reports in *New Scientist*, 1992.

Chapter 4

The *Proceedings of the Ecological Society of Australia*, volume 16 (1990), provides several articles on the impact of Europeans on Australia's vegetation. A good summary of European impact can be found in the chapter by Adamson and Fox in Smith (1982) *op. cit.*

A continent-wide overview of spatial vegetation change since European settlement can be found in AUSLIG (1990), Volume 6 Vegetation, *Atlas of Australia's Resources*, AGPS, Canberra.

The transformation of the plains country is reviewed in Kirkpatrick, J.B., Gilfedder, L. and Fensham, R.J. (1988), *City Parks and Cemeteries*, Tasmanian Conservation Trust, Hobart.

Information on tree farming in Australia can be accessed in the various reports of the Resource Assessment Commission's Forest and Timber Inquiry and the Report of the Plantations Advisory Council, all of which were produced by the AGPS in 1991 and 1992.

Data on the reservation status of species and communities was obtained from Amos et al. (1993) op. cit.

Chapter 5

Resource Assessment Commission (1992), *Forestry and Timber Inquiry—Final Report*, AGPS, Canberra, provides a comprehensive overview of all aspects of forestry and timber production.

Kirkpatrick, J.B., Meredith, C., Norton, T. and Fensham, R.J. (1990), *The Ecological Future of Australia's Forests*, Australian Conservation Foundation, Melbourne, discusses in depth most aspects of the interaction between forest use and forest ecology.

The Mt Lofty Range example of relative drought resistance was drawn from Sinclair, R. (1980), Water potential and stomatal conductance of three *Eucalyptus* species in the Mt Lofty Ranges, South Australia, *Australian Journal of Botany* 28, 499–510.

Chapter 6

Gill et al. (1981) op. cit. is a comprehensive treatment of the impact of fire on the Australian biota. Kirkpatrick, J.B. (1991b), *Tasmania's Native Bush—A Management Handbook*, TEC, Hobart, provides details of the ecological impact of fire, and appropriate fire management techniques for several ecosystem types widespread in Australia. Harrington, G.N., Wilson, A.D. and Young, M.D. (1984), *Management of Australia's Rangelands*, CSIRO, Melbourne, provides details of fire ecology and fire management for the ecosystems that cover most of the continent.

Chapter 7

Humphries, S.E., Groves, R.H. and Mitchell, D.S. (1992), *Plant Invasions of Australian Ecosystems*, ANPWS, Canberra, provides a comprehensive and detailed review of the subject matter covered by this chapter.

The data on Rodondo and Curtis Islands is from the *Papers and Proceedings of the Royal Society of Tasmania*, volume 108, 1974.

Specht, R.L. (1974), *Vegetation of South Australia*, Government Printer, Adelaide, provided the data on the nutrient levels of various wastes.

Buchanan, R.A. (1989), *Bush Regeneration*, TAFE Student Learning Publ., Sydney, and Kirkpatrick (1991b) op cit. give details of various methods of weed control.

Harrington et al. (1984) op. cit. discuss grazing management and the ecological impacts of animals.

Recent papers on the impact of grazing on Australian vegetation include: Leigh, J.H. et al. (1987), Effects of rabbit grazing and fire on a subalpine environment. I. Herbaceous and shrubby vegetation, *Australian Journal of Botany* 35,

433–64; Williams, J.H. and Ashton, D.H. (1987), Effects of disturbance and grazing by cattle on the dynamics of heathland and grassland communities on the Bogong High Plains, *Australian Journal of Botany* 35, 413–31; Gibson, N. and Kirkpatrick, J.B. (1989), Effects of cessation of grazing on the grasslands and grassy woodlands of the Central Plateau, Tasmania, *Australian Journal of Botany* 37, 55–63.

Chapter 8

Amos *et al.* (1993) *op. cit.* provides a review of the theoretical base for conservation planning. More details of methods can be found in Margules, C.R. and Austin, M.P. (1991), *Nature Conservation: Cost Effective Biological Surveys and Data Analysis*, CSIRO, Melbourne.

Groves, R.H. (1981) *op. cit.* covers the ecology of Australia's major vegetation types. A revised edition of this book is due soon.

Humphries *et al.* (1992) *op. cit.* provides more detail on threatening exotic plants.

Harrington *et al.* (1984) *op. cit.* and Kirkpatrick (1991b) *op. cit.* provide further analysis of the major management problems for Australian vegetation types.

Index

(Common plant names, when used, are cross-referenced with scientific plant names. These follow Hnatiuk (see bibliography))

Aborigines (see also Tiwi people)
  arrival in Australia   28–30
  current effects on environment   35
  nature of prehistoric impact   30–4
  relationship to nature   34–5
  reserve management   80
  use of fire   30, 56–9, 63, 65
Acacia (wattles)
  adaptations to water stress   12
  effect of Cinnamon Fungus on   93–4
  effect of rabbits on   97
  impact of fire on   74, 76, 94
  impact of grazing on   76, 87
  regeneration of   17, 19, 31, 74
  weed invasions into   92
  weed species of   117
Acacia aneura (Mulga)   97
Acacia nilotica   119
Acacia sophorae (Coast Wattle)   19, 91
Acacia suaveolens   12
acclimatisation societies   95
agriculture
  agricultural ecosystem   4–49
  biodiversity and   3
  integrated land management for   120
  land clearance for   37–41, 99, 117–19
  organic   48–9
  permaculture   49
  role of CSIRO in development of   43–4
  salinisation   62
allelopathy   17
aliens (see exotics)
Allocasuarina   12, 29
alliance   100
alpine vegetation   52, 69–71, 76–9, 94, 111, 116
Ammophila arenaria (Marram Grass)   85, 120
ants   17
Araucaria bidwillii (Bunya Pine)   31
Araucaria cunninghamii (Hoop Pine)   63
Araucaria heterophylla (Norfolk Island Pine)   49
Ardrey, Robert   32

Ashton, David 33
Asteraceae (composites, daisies) 27
Astrebla (Mitchell) grasslands 118–19
Atherosperma moschatum (Southern Sassafras) 58, 64
Athrotaxis cupressoides (Pencil Pine) 69, 77
Athrotaxis selaginoides (King Billy Pine) 64, 69
atmospheric factor 14
Australian Conservation Foundation 4
Austrofestuca littoralis 19
bandicoot 18
Banksia 93–4
Barrow Island 104
Bathurst Harbour 92
Bathurst Island 35
beta diversity 23
Big Scrub 39
biodiversity (see biological diversity)
biological control 90, 1
biological diversity 1–3, 99
  economic worth of 3
biotic factor 16–17
bioregions 120
biotic realms 82
Bitou Bush (see *Chrysanthemoides monilifera*)
Blackberry (see *Rubus fruticosus*)
Blackbird 88
Black Rock 48
Blue Thunbergia (see *Thunbergia grandiflora*)
Bogong High Plains 76
Boneseed (see *Chrysanthemoides monilifera*)
Borya 12
Brachiara mutica 119
Bracken (see *Pteridium esculentum*)
Bradley method 90
Brigalow 40, 51
Broom (see *Genista monspessulana*)
buffalo 35, 95
Bunya Pine (see *Araucaria bidwillii*)
bush peas (see Fabaceae)
Buttongrass (see *Gymnoschoenus sphaerocephalus*)
Californian Redwood (see *Sequoia sempervirens*)
Callitris (Cypress Pine) 15, 30, 65
camel 95
camping, impact of 115
Cannabis sativa (Marijuana) 31
Carpobrotus rossii (Pigface) 15
cassava 31
cat 33, 95
catchment values 78–9, 108
Catsear (see *Hypchaeris radicata*)
cattle 16, 44, 75–9, 95
Cenchrus ciliaris 119
Central Plateau 77, 101
charcoal, use in vegetation history 26, 29, 70
Chrysanthemoides monilifera (Boneseed, South African Boneseed, Bitou Bush) 79, 88, 89, 91, 120
Cinnamon Fungus (see *Phytophthora cinnamomi*)
classification 7–8, 21–2, 100–2
  community 21–2, 100–2
  general 7
  species 7–8
clearfelling 57
climatic change 25–6, 29, 66, 67, 106–8, 116
clinal variation 10
coastal vegetation 119–20
coast disease 44
Coast Wattle (see *Acacia sophorae*)
Commonwealth Scientific and Industrial Research Organisation (CSIRO) 43–4, 76, 91, 98
community definition 21–2, 100
competitors 16, 74
composites (see Asteraceae)
Conebush (see *Isopogon ceratophyllus*)
coniferous heath 69–72
Coonalpyn Downs 44
Cotoneaster (Cotoneaster) 88
Cradle Mt.-Lake St. Clair National Park 113
Crassula 12
crassulacean metabolism 12
Crataegus monogyna (Hawthorn) 88
Creeping Pine (see *Microcachrys tetragona*)
crops 40, 41, 48–9

Cryptostegia grandiflora 116
cultural vegetation 48–50
Cupressus macrocarpa (Monterey Cypress) 87
Curr, Edmund 30
Cyperaceae (Sedge) 71
Cypress Pine (see *Callitris*)
Dandenong Ranges 33
daisies (see Asteraceae)
Danthonia (Wallaby Grass) 42
dating of deposits 28–9
dating of trees 97
Deciduous Beech (see *Nothofagus gunnii*)
deciduous heath 69–72
desert vegetation 76, 97, 111, 119
Dingo 32, 34–5
disjunctions 11
dispersal
  ecology of 19–20, 84–5
  of alpine conifers 71
  of *Chrysanthemoides* 91
  of monsoon forest trees 65
  of *Phytophthora* 92–5
  of *Pittosporum* 88
  speed of 85, 87
disturbance
  and exotic plants 84–5
  animal 18
  digging 32
  endogenous 18, 62
  exogenous 18, 62
  intermediate 22, 65
  urban 49–50
diversity 22–3
dog 28, 33, 34
drought 13, 61
dry eucalypt (sclerophyll) forest 117
Echinochloa polystachia 119
ecological horticulture 50
ecological reserves 100
ecological response curves 12, 105–6
ecological services 2
ecologically sustainable development 3–4
ecological theory 4–5
ecosystem 20–1
ecotypic variation 10
edaphic factor

  effect of forestry on soils 58–9, 63
  effect on plants 15
  effect of trace element deficiencies 44
  relationship to exotic invasion 84–7, 93–4
  soil nutrient balance 21
edge effects 108
Eichhornia crassipes 119
endemism 11, 106–8
environmental domain analysis 101
environmental weeds 82
Epacridaceae (southern heaths) 18, 93
Errindundra Plateau 58
Eucalyptus (see also entries under individual species and vegetation)
  conservation of genetic variation 47
  dispersal of 20
  distribution in relation to understorey 101
  effects of *Phytophthora* on 93
  effects of temperature on germination of 17–18
  fire ecology of 16, 72, 76–7
  impact of logging on 55–62, 65
  light ecology of 14
  resistance to clearing 40–1
  rural dieback of 42
Eucalyptus delegatensis 8, 12
Eucalyptus gigantea 8
Eucalyptus globulus (Tasmanian Blue Gum) 7, 47
Eucalyptus marginata (Jarrah) 93
Eucalyptus nitens (Shining Gum) 47, 58
Eucalyptus obliqua (Messmate Stringybark) 60
Eucalyptus pauciflora (Snow Gum) 76–7, 118
Eucalyptus pauciflora (Poplar Box) 40
Eucalyptus regnans (Mountain Ash) 8, 33, 87
Eucalyptus sieberana 8
Eucalyptus sieberi 8
Eucalyptus tenuiramis 8
Eucryphia lucida (Leatherwood) 64

euro 104
European Wasp 95
evolution 8–10
extinction 10–11, 22, 32, 67, 103
fauna
   acclimatisation societies 95
   and disturbance 18
   hunting of 32
   impacts of grazing (see 'grazing' and 'stock, impacts of')
   introduced species 16, 22, 25, 28–9, 33–5, 39, 41–4, 52, 67, 69, 75–9, 88, 95, 97–8, 112, 116, 118–19
   native species 2, 16–18, 32–5, 66, 90–1, 104, 110
Fabaceae (bush peas, native peas) 15, 74, 93
Federation Peak 101
Fern-leaved Mahonia (see *Mahonia lomariifolia*)
fire
   and grazing 75–9
   and protected areas 79–81
   and alpine vegetation 69–71
   and rainforest 63–5
   and tree establishement 42, 56–9
   and weeds 79, 87
   ecology of 15–16
   effect of seasonality 74–5
   effect on flammability 21
   history of 68–9
   management 71–4, 79–81
   prehistory of 29–30, 68
Flatweed (see *Hypochaeris radicata*)
Flinders Chase National Park 79
Flinders Island 95
flora (see also entries for individual taxa)
   fire and 15–16, 68–81
   introduced species 11, 31, 44–7, 49, 79, 82–98, 116, 118–120
   rare and threatened species 102–3
forest 25, 31, 35, 54–67, 116–17
   clearance 39–40
   history 25, 54–5
   nutrient cycling 58–9
forestry

   impact on eucalypt forest 55–62, 66–7
   impact on forests of wet/dry tropics 65
   impact on rainforest 62–5
   philosophy 55
   plantations 45–7
   silviculture 55–6, 62–3
frost 42
fuel reduction burning 72–4, 79
fundamental niche 18, 23
fynbosch 22
garden fashions 49
gardens 49–50
gathering 31
Genista monspessulana (broom) 79
genocide 35
glacial periods 25–6, 29, 106–7
Glyceria maxima 119
goat 95
Gondwana 25
gorse (see *Ulex europaeus*)
Gott, Beth 31
grasses (see Poaceae)
grassland
   clearance and degradation of 35, 39, 41–3
   fire ecology of 30, 76–8
   prospects for 118–19
grass-tree (see *Xanthorrhoea*)
Gymnoschoenus sphaerocephalus (buttongrass) 71, 93
grassy woodland 41–3, 108, 118
grazing (see also stock)
   and fire 69, 75–9
   area affected by 39
   impact on native vegetation 41–3, 95, 116, 118–19
group selection 56
Hattah-Kulkyne National Park 2
Hawthorn (see *Crataegus monogyna*)
hazard reduction burning (see fuel reduction burning)
heath 40, 44–5, 51, 86, 94, 110
Helichrysum hookeri (Kerosene Bush) 77
Helichrysum paralium 19
herbicides 89
Himalayan Honeysuckle (see *Leycesteria formosa*)

Hogan Group 83–4
Holcus lanatus (Yorkshire Fog Grass) 85
Holocene 25, 27
Hoop Pine (see *Araucaria cunninghamii*)
Hunting 32
Huon Pine (see *Lagarostrobus franklinii*)
hybridisation 10, 60
Hymenachne amplexicaulis 119
Hypochaeris radicata (Catsear, Flatweed) 49, 85
integrated land management 120
introduced species (see under flora and fauna)
inundation 38–9
island biogeography 20, 105–6
Isopogon ceratophyllus (Cone Bush) 94
Jarrah (see *Eucalyptus marginata*)
Jones, Rhys 28, 34
Kakadu National Park 75
kangaroo apple (see *Solanum laciniatum and S. vescum*)
Kangaroo Grass (see *Themeda triandra*)
Kangaroo Island 79, 95
kangaroos 2, 110
Kerosene Bush (see *Helichrysum hookeri*)
Kershaw, Peter 29
King Billy Pine (see *Athrotaxis selaginoides*)
King Island Orange-bellied Parrot on 104
Koala 16
Koonamore 95, 97
Kosciusko 76–8
kwongan 22
Lagarostrobus franklinii (Huon Pine) 20, 64
Lake George 29
land clearance 37–41, 43–7, 99, 117–19
Lantana camara (Lantana) 83
Leatherwood (see *Eucryphia lucida*) 64
Leucaena leucocephala 44
Leycesteria formosa (Himalayan Honeysuckle) 83
light factor 14, 56
Ligustrum (privet) 88
Liliaceae (lilies) 15
limits of tolerance 11–17
logging 39, 53–67
Lyrebird, Superb 33
Macassan traders 28
macrofossils 26–7
Macrozamia 33
Mahonia lomariifolia (Fern-leaved Mahonia) 49
mallee 12, 40, 79, 97, 118
mangroves 108, 120
Marijuana (see *Cannabis sativa*)
Marram Grass (see *Ammophila arenaria*)
Megafauna 32–3, 66
Melaleuca 14
Messmate Stringybark (see *Eucalyptus obliqua*)
metapopulations 104
Microcachrys tetragona (Creeping Pine) 71
Microseris lanceolata (Myrniong) 31
Microstrobus fitzgeraldii 71
Mimosa pigra 87, 119
Mimosa pudica (Sensitive Plant) 49
minimum set techniques 109
minimum viable population 103
Mission Grass (see *Pennisetum polystachion*)
Mitchell grassland (See *Astrebla grassland*) 118–19
mixed forest 58
Monterey Cypress (see *Cupressus macrocarpa*)
Monterey Pine (see *Pinus radiata*)
Mountain Ash (see *Eucalyptus regnans*)
Mountain Plum Pine (see *Podocarpus lawrencei*)
Mount Lofty Range 60
Mueller, Baron von 83
Mulga (see *Acacia aneura*)
multiple use 55
Murray River 39
Myrniong (see *Microseris lanceolata*)
Myrtle Beech (see *Nothofagus cunninghamii*)

National Forest Policy   67
National Parks   44, 50–2, 58, 62, 79–81, 99–116
native peas (see Fabaceae)
New England Tablelands   42
Ninety Mile Desert/Plain   44, 86
Nix, Henry   101
Norfolk Island Pine (see *Araucaria heterophylla*)
Nothofagus   27
Nothofagus cunninghamii (Myrtle Beech)   64–5
Nothofagus gunnii (Deciduous Beech)   16, 30, 69
Nullarbor Plain   20
Ocean Beach   46
off-road vehicles   110–11
old growth forest   66
Olea europaea (Olive)   88
Onion-grass (see *Romulea rosea*)
open-forest (see dry eucalypt forest)
Opuntia (pickly pear)   91
Orange-bellied Parrot   104
Orchidaceae (orchids)   15, 17
organic agriculture   48–9
Overland Track   113
Oxalis corniculata (Soursob)   83
palynology   26–8, 70
pastoralism (see stock, impacts of)
Pandanus   33
peas (see Fabaceae)
Pencil Pine (see *Athrotaxis cupressoides*)
Pennisetum polystachion (Mission Grass)   117
permaculture   49
peat   63
Phytophthora cinnamomi (Cinnamon Fungus)   67, 79, 92–5, 118, 119
pig   95
Pigface (see *Carpobrotus rossii*)
Pilbara   76
Pink Swamp Heath (see *Sprengelia incarnata*)
Pinus radiata (Monterey Pine, Radiata Pine)   45–7
Pittosporum undulatum (Sweet Pittosporum)   88
plantations   35, 45–7
Plechtrachne (Spinifex)   16, 76

Poa spp (Snow Grass)   41, 77–8
Poaceae (grasses)   16, 74
Podocarpus lawrencei (Mountain Plum Pine)   71
pollen records   26–8, 70
   interpretation of   27
Poplar Box (See *Eucalyptus populnea*)   40
population viability analysis   104
predation   16
Prickly Pear (see *Opuntia*)
primary species   63
privet (see *Ligustrum*)
protected areas (see National Parks)
Pteridium esculentum (Bracken)   15, 31, 41, 74, 82
Quaternary   22, 25, 29, 67
rabbit   35, 95, 97–8, 118
Radiata Pine (see *Pinus radiata*)
rainforest
   biotic interactions within   17
   changes in extent   11, 30, 39
   effect of *Phytophthora* on   93
   fire ecology of   15, 30, 56
   fire management of   71–2
   impacts of logging on   54, 62–5
   prospects for   116
   reservation of   51
rare and threatened species   102–3
reclamation   39
recreation
   camping, impact of   115
   gardening   49–50
   hazard reduction burning   73
   National Parks   44, 50–2, 58, 62, 79–81, 99–116
   off-road vehicles   110–11
   ski development   112
   tourist development   112
   walking tracks, impact of   112–16
Red Cedar (see *Toona australis*)
Red Kangaroo   104
refugia   66, 106–7
regeneration niche   17–18
relicts   11
reserve management   110–16
reserve planning   103–9
reserves (see National Parks)
reserve selection techniques   109
Resource Assessment Commission   66

rhododendron 49
Richea scoparia (scoparia) 101
rights of living world 1–2
Rodondo Island, exotics on 83–4
Romulea rosea (Onion-grass) 90
Rosa spp. (rose) 49
Rubus fruticosus (Blackberry) 83, 89
ruderals 16, 19, 49
rural dieback 42
Rutidosis leptorrynchoides 43
salinisation 62
Salix spp. (willow) 87
saltbush shrublands 51
saltmarsh 104, 108, 110, 120
Salvinia molesta 119
sand mining 35
sassafras (see *Atherosperma moschatum*)
savannah woodland (see grassy woodland)
Schismus barbatus 95
Scoparia (see *Richea scoparia*)
Sea Rocket (see *Cakile*)
secondary species 62–3
sedge (see Cyperaceae)
Sensitive Plant (see *Mimosa pudica*)
Sequoia sempervirens (Californian Redwood) 87
sheep 44, 75–9
Sherbrooke Forest 33–4
Shining Gum (see *Eucalyptus nitens*)
silvicultural systems 55–8
Singh, Gurdip 29
ski development 112
Snow Grass (see *Poa*)
Snow Gum (see *Eucalyptus pauciflora*)
soil erosion 59, 74, 78, 111, 113, 115
soil factor (see edaphic factor)
Solanum laciniatum (kangaroo apple) 31
Solanum vescum (kangaroo apple) 31
Sorghum 75
Soursob (see *Oxalis corniculata*)
South African Boneseed (see *Chrysanthemoides monilifera*)
South Gippsland Hills 39
southern heath (see Epacridaceae)
South West National Park 104

sparrows 16
Spartina townsendii 120
Spear Grass (see *Stipa*)
Specht, Ray 21, 100
speciation 9–10, 22
 allopatric 9, 22
 parapatric 9
 sympatric 9
species richness 22–3, 49, 62
Sphagnum 78
spinifex (see *Triodia and Plectrachne*)
Spinifex sericeus 19
spiritual values of living world 2
Sprengelia incarnata (Pink Swamp Heath) 94
Spur Velleia (see *Velliea paradoxa*)
State Forest 54–5, 62, 64
Stipa (Spear Grass) 42, 95, 118
Stipa neesiana 118
St Kilda 50
stock (see also grazing)
 area affected by 39
 impact on alpine vegetation 52, 76–9, 116
 impact on coastal vegetation 120
 impact on desert vegetation 76, 95, 119
 impact on dry eucalypt forest 117
 impact on grassy vegetation 41–3, 118–19
 interaction with fire 69, 75–9
succession 18–19, 56–7, 67, 77
sustainable use 55
sustained yield 55
Sweet Pittosporum (see *Pittosporum undulatum*)
tali open-forest (see wet eucalypt forest)
Tamarix aphylla 119
Tasmanian Devil 32, 35
Tasmanian Forestry Commission 64
temperature factor 13–14, 77, 93
Tetrarrhena juncea (wiregrass) 33
Themeda triandra (kangaroo grass) 42, 77
theory of island biogeography 20, 105–6
Thunbergia grandiflora (Blue Thunbergia) 49, 116

Thylacine 32, 35
Tiwi people 35
tolerators 16
Toona australis (Red Cedar) 54
tourist development 112
trace element deficiencies 44
treeline 13
Triodia spp. (spinifex) 16, 76
tropical eucalypt forest 117
trout 112
Ulex europaeus (Gorse) 79, 83, 85, 89, 118
Uluru 112
urbanisation, impact of 47–8
urban bush 47–8, 86, 88–9
vegetation
   alpine 52, 69–71, 76–9, 94, 111, 116
   Astrebla (Mitchell) grasslands 118–19
   brigalow 40, 51
   coastal 119–20
   cultural 48–50
   desert 76, 97, 111, 119
   dry eucalypt (sclerophyll) forest 117
   fynbosch 22
   grassland 30, 35, 39, 41–3, 51, 76, 77–8, 118–19
   grassy woodland 41–3, 108, 118
   heath 40, 44–5, 51, 86, 94, 110
   kwongan 22
   mallee 12, 40, 79, 97, 118
   mangroves 108, 120
   mixed forest 58
   rainforest 11, 15, 17, 30, 39, 51, 54, 56, 62–5, 71, 88, 93, 101, 116

saltbush shrublands 51
saltmarsh 104, 108, 110, 120
tropical eucalypt forest 117
urban bush 47–8, 86, 88–9
Wallum 86
wet eucalypt forest 8, 33, 58, 87, 116–17
wetlands 31, 38, 108, 119
vegetation dynamics 18–19, 56–7, 67, 77
Velleia paradoxa (Spur Velleia) 16
walking tracks, impact of 112–16
wallaby 33
Wallaby Grass (see *Danthonia*)
Wallum 86
water factor 12, 38, 60–1, 78–9
wattles (see *Acacia*)
Wedge-tailed Eagle 16
weeds (see also ruderals) 82
wet eucalypt forest 8, 33, 58, 87, 116–17
wetlands 31, 38, 108, 119
wet sclerophyll forest (see wet eucalypt forest)
wilderness 55, 91, 109
willow (see *Salix*)
Wilsons Promontory National Park 79, 94, 114
Wire Grass (see *Tetrarrhena juncea*)
wombat 33
Woolnorth 45
World Commission on Environment and Development 4
Xanthorrhoea spp. (grass-tree) 94–5
Yorkshire Fog Grass (see *Holcus lanatus*)